Concepts of War, 1650–1900

Value Inquiry Book Series

VOLUME 383

Studies in the History of Western Philosophy

The titles published in this series are listed at *brill.com/vibs* and *brill.com/shwp*

Concepts of War, 1650–1900

From Free-Rider Strategies to Survival of the Fittest

By

Paul Schuurman

BRILL

LEIDEN | BOSTON

Cover illustration: The Battle of Marengo (1800) by Louis-François Lejeune. Painting in the public domain.

Library of Congress Cataloging-in-Publication Data

Names: Schuurman, Paul, author.
Title: Concepts of war, 1650-1900 : from free-rider strategies to survival of the fittest / by Paul Schuurman.
Other titles: From free-rider strategies to survival of the fittest
Description: Leiden ; Boston : Brill, 2023. | Series: Value inquiry book series, 0929-8436 ; 383 | Includes bibliographical references and index. | Summary: "This book discuss the key concepts that philosophers and generals of this era developed to grasp and influence the dramatic phenomenon of war"– Provided by publisher.
Identifiers: LCCN 2023002126 (print) | LCCN 2023002127 (ebook) | ISBN 9789004536661 (hardback) | ISBN 9789004536678 (ebook)
Subjects: LCSH: War (Philosophy) | War.
Classification: LCC U21.2 .S34 2023 (print) | LCC U21.2 (ebook) | DDC 172/.42–dc23/eng/20230118
LC record available at https://lccn.loc.gov/2023002126
LC ebook record available at https://lccn.loc.gov/2023002127

Typeface for the Latin, Greek, and Cyrillic scripts: "Brill". See and download: brill.com/brill-typeface.

ISSN 0929-8436
ISBN 978-90-04-53666-1 (hardback)
ISBN 978-90-04-53667-8 (e-book)

This book is printed on acid-free paper and produced in a sustainable manner

For Sascha Brouwer, Feline van der Gun, Nora van der Gun, Gaïa Landolsi, Eileen Landolsi, Jad Landolsi, Noa Ooijkaas, Boas Ooijkaas, Teun Rustenburg, Pelle Rustenburg, Birk Rustenburg, Ezra Rustenburg, Serra van der Vegte, and Evelynn Zevenbergen—may they prosper in peace

∵

Contents

Preface

The present collection of eight essays on concepts of war in the history of ideas is based on previously published articles that I have corrected, updated, and expanded. Some essays have been provided with a new title.[1] English translations of foreign quotations are by me unless stated otherwise in the footnotes.

Thanks are due to Samir Azrioual, Annie Barendregt, Jasper Blaas, Pieter Boots, Erwin van den Bosch, Simona Brolsma, the late Hans Brouwer, Mark Brummel, Wiep van Bunge, Bert Claessen, Floran Clopin, Paul Donker, Erno Eekelschot, Ivo Geradts, Paul van der Gun, Knud Haakonssen, Nico Heidari Tari, Stéphanie Helfferich, Joost Hengstmengel, Andreas Herberg-Rothe, Cees van Houten, Hu Junhao, Ron Huising, Youssef Jemmad, Alexander Kiel, Abdessatar Landolsi, Charissa de Lange, Bart Leeuwenburgh, Ma Dengfeng, Li Zhengting, Robert van de Medevoort, Tim de Mey, Miyagawa Hitomi, Debby Mudde, Fred Muller, Olga Oleksiuk, Priscilla Pheijfer, Maria Quartero, the late John Rogers, Johannes Rustenburg, Marieke Rustenburg, Sassan Sangsari, Margreet Scheurwater, Bart Schoonens, Marius Schuurman, Corella Schuurman, Leon Schuurman, Dennis Schuurman, Mark Siekerman, Bram Snepvangers, Betty Stad, Elena Ulanova, Michel van de Vegte, Mart Verhage, Catherine Volpilhac-Auger, Irene Wesdorp, Richard Whatmore, the late Michiel Wielema, Andrew Woodward, and Yang Lo-Hsuan.

1 The original articles (with the numbers between square brackets corresponding with the chapters of this book) are: [1] 'The Cat's Grand Strategy. Pieter de la Court (1618–1685) on Holland and the Challenges and Prospects of Free-Riding Behaviour During the General Crisis Of the Seventeenth Century', *History of European Ideas*, 40 (2014), 1–19. http://dx.doi.org/10.1080/01916599.2014.965783; [2] 'Fénelon on Luxury, War and Trade in the *Telemachus*', *History of European Ideas*, 38 (2012), 1–21; http://dx.doi.org/10.1016/j.histeuroideas.2011.07.013; [3] 'Determinism and Causal Feedback Loops in Montesquieu's Explanation for the Military Rise and Fall of Rome', *British Journal for the History of Philosophy*, 21 (2013), 507–28. http://dx.doi.org/10.1080/09608788.2013.771612; [4] 'War as a System: A Three-Stage Model for the Development of Clausewitz's Thinking on Military Conflict and Its Constraints', *Journal of Strategic Studies*, 37 (2014), 926–48. http://dx.doi.org/10.1080/01402390.2014.933316; [5] 'What-If at Waterlo: Carl von Clausewitz's Use of Historical Counterfactuals in his History of the Campaign of 1815', *Journal of Strategic Studies*, 40 (2017), 1016–1038. https://doi.org/10.1080/01402390.2017.1308862; [6] 'Models of War 1770–1830: The Birth of Wargames and the Trade-off Between Realism and Simplicity', *History of European Ideas*, 43 (2017), 442–45. http://dx.doi.org/10.1080/01916599.2017.1366928; [7] 'A Game of Contexts: Prussian-German Professional Wargames and the Leadership Concepts of Mission Tactics 1870–1880', *War in History*, 28 (2021), 504–24. https://doi.org/10.1177/0968344519855104; [8] 'Herbert Spencer and the Paradox of War', *Intellectual History Review*, 26 (2016), 519–35. http://dx.doi.org/10.1080/17496977.2016.1220734. I would like to thank these journals for their permission to use my articles for this book.

Figures and Tables

Figures

Tables

Introduction

From the earliest beginnings of civilization war has been a potent and dynamic driver of history. For example, many of the earliest city-states were the product of war. Early city-states were religious and civil centres, with marketplaces and artisan workshops contributing to their development. Yet remarkably enough up to 80 per cent of early city-state dwellers were farmers. Accompanied by their cattle, they left their city-state at dawn and returned at dusk. Farmers took the crowded living conditions and the increased risk of epidemic diseases for granted, because cities provided security against raids. City-states not only provided protection; they were also used as a base for offensive operations. In this way one city-state swiftly stimulated the birth of other city-states. Hence, we typically see them popping up one shortly after the other in clusters of dozens or even hundreds in Mesopotamia, Ancient Greece, the Valley of Mexico, and mediaeval northern Italy. In all these cases warfare prompted defensive coalescence, bringing urbanization and the trappings of civilization in its wake.[1]

In the period covered by the eight essays in this book, 1650–1900, warfare remained an engine of change. The decades around 1650 saw a disastrous global wave of climate change, famine, epidemics, and warfare. During the so-called General Crisis of the Seventeenth Century wars peaked in frequency and intensity, devastating entire provinces in Central Europe. In subsequent decades armies continued to make ever increasing demands in manpower, equipment, ammunition, and food. In order to pay for these resources states modernized their institutions, thus boosting their ability to raise money by taxes and deficit financing. War made states and Prussia was said to be an army with a state rather than the other way around. Commerce and manufacture were often impeded by actual warfare, but at the same time their potential was harnessed to military ends.

In the eighteenth century, European competition on an increasingly global scale caused strains on national budgets and contributed to bankruptcy and revolution in France. Political revolution brought fundamental changes in economic, social, political, and military institutions. These changes increased military potential. Increased military effectiveness combined with ideological fervour submerged Europe in warfare. Fundamental reforms in a whole range of sectors in Napoleonic Europe were undertaken with the aim of boosting the war machine of the French Empire. War itself was revolutionized by changes on the tactical and higher operational levels, forcing all European powers to adapt

1 Gat, *War*, 269–78.

 | DOI:10.1163/9789004536678_002

with similar reforms. In the decades following political revolution and revolutionary warfare, industrial development gathered momentum. Demographic growth, increased firepower, railways, and telegraphs continued to change the face of war. Renewed warfare in the middle of the nineteenth-century brought defeat to one empire and the complete collapse of another, while freshly uniting two other nations. By 1900 all the elements for the poisonous cocktail of industrial total warfare were firmly in place.

Given the dramatic importance of war, it is not surprising that some of the best minds between 1650 and 1900 applied existing concepts and developed new concepts to grasp the nature of warfare. Moreover, all assumed that their ideas and concepts could actually influence the robustly factual and bloody reality of war. Even the most determinist and the least voluntaristic inclined of these thinkers on war used their analysis to propose policies, strategies, and tactics for the pursuit of military victory. Also, sometimes at the same time, they proposed measures that should lead to peace and the abolition of war. Pieter de la Court (1618–1685) recommended a commercial grand strategy for Holland. François Fénelon (1651–1715) advocated a pacifist agricultural policy that should simultaneously strengthen the military potential of France. Montesquieu (1689–1755) argued for a vigorous national policy that would be inspired by commercial Carthage rather than belligerent Rome. Carl von Clausewitz (1780–1831) proposed a form of military instruction that should stimulate autonomous thinking rather than install static rules of conduct that would prove useless in the turbulence of battle. Designers of wargames hoped that their games would prepare officers for the real game of warfare itself. And although for Herbert Spencer (1820–1903) human history was only one part of a vast and completely determined cosmic evolution, he still allowed himself the luxury of advocating a minimum of state interference as the best guarantee against the twin spectres of war and socialism.

So, between 1650 and 1900 there was a wide-spread appreciation of the importance of war. Joseph de Maistre (1753–1821) observed that 'War is a great and profound subject which concerns the philosopher as much as the general'.[2] Indeed, in the eighteenth century we see preciously little separation between the military and the civil spheres, especially at the upper strata of society. Military and intellectual pursuits were easily combined. Arguably the two greatest generals between 1650 and 1900, Napoleon and Frederick the Great of Prussia, provide good examples of this point. The flute-playing Prussian king composed instrumental and vocal music while also trying his

2 Quoted in Bell, *The First Total War*, 52; see also Rosenberg, 'Joseph de Maistre', 43–7.

hand at fiction, history, political philosophy, and a range of military topics. The young Napoleon wrote both fiction and history. He remained a voracious reader for the rest of his life. After his fall from power, he devoted many pages of his *Mémorial* to a disgruntled analysis of the causes for his military defeats.

The nineteenth century saw an increased separation between the military and the civil sphere. Yet several intellectual developments helped to maintain an interest in war, not only by officers but also by civilians. Hegel, Nietzsche, and other philosophers emphasised the functional and regenerating blessings of warfare. Darwin's biological theory of evolution by natural selection was used by many to stress the importance of a military struggle for survival between nations. Besides this, history in the nineteenth century was still a predominantly literary and narrative art that stressed the agency of a few 'great men'; no story was more dramatic, and no subject more compelling than a military campaign crowned by a climactic battle.

These appreciations changed in the course of the twentieth century. Marxist historians stressed the economic struggle between classes rather than the military struggle between nations. Many Marxists held that the latter was used by leading capitalists to deflect the working classes from the former. Marxist influences were also present in the influential French historical school gathered around the journal *Annales d'histoire économique et sociale* (1929), which downplayed *histoire événementielle* and focussed on deeper geological, social, and economic strata. Military battles were considered as mere spume on the waves of history. Furthermore, liberal historians and philosophers after the Second World War increasingly came to regard war as primitive, irrational, and alien to modern civilization. The collapse of the Soviet Union and the end of the Cold War seemed to confirm the irrelevance of war for our times. In a highly influential article of 1989 (followed by a book with the same title) Francis Fukuyama noted an unabashed victory of economic and political liberalism, an end of ideological and military conflict, and hence *the end of history*.[3]

Just when the Western world leaned back to enjoy the dividends of peace at the start of the twenty-first century, the terror attack against the Twin Towers forced a renewed assessment of the reality and relevance of violent conflict. This prompted renewed attention for an article (and subsequent book) which had been published already in 1993 by Samuel Huntington, 'The Clash of Civilizations'.[4] In this article, written in reaction to Fukuyama, it was argued that military conflict is here to stay, and that it will first of all be driven by

3 Fukuyama, 'The End of History', 3–18; see also Fukuyama, *The End of History*.

4 Huntington, 'The Clash of Civilizations', 22–49; see also, Huntington, *The Clash of Civilizations and the Remaking of the World Order*.

cultural and religious differences. Huntington predicted that the fault lines between civilizations will be the battle lines of the future. Moreover, the combined effect of a depletion of natural resources, aggravated by a disastrously long list of ecological problems, is widely regarded as a trigger for future conflicts.[5] Finally, the instability associated with the decline of old superpowers and the rise of a new contender, has contributed to a more urgent attention to the spectre of war.

The changing vicissitudes of war as a topic of scholarly attention may have resonated in the reception given to a Master's course on war in the history of ideas that I introduced in 2011 at the philosophy department of my university. This course formed my research agenda for the subsequent decade. My initiative received friendly comments, but eyebrows were raised when my reproductions of great generals and the splendid maps with red and blue arrows of Westpoint Academy started to spill over from the interior walls of my office to the corridor of my department. More fundamental was the question raised by another colleague: why should my course be offered in a philosophy department? The question implied doubt about war as a 'great and profound' topic in the history of ideas. The present collection is my attempt to answer this question.

I present a personal selection of key concepts used by some very famous and some rather less famous intellectuals. These concepts were extremely diverse, but they all helped to make sense of the complexities of war in a systematic, fundamental, and multidisciplinary way. De la Court used the game-theoretical concept of *free-riding* in order to grasp the grand-strategic challenges and opportunities for Holland during the General Crisis of the Seventeenth Century. Fénelon's spiritual concept of *quietism* influenced his utopian recommendations for the interrelated topics of luxury, war, and trade. Montesquieu's deterministic explanations for the military rise and fall of Rome was informed by the system theoretical concepts of *positive and negative feedback loops*. In the first chapter on Clausewitz we see how he used the same conceptual devices for an elegant three-stage interpretation of the constraints on warfare. The second chapter on Clausewitz discusses his use of *historical counterfactuals* in his analysis of the Waterloo Campaign of 1815. The chapter on professional wargames between 1770–1815 as *models* discusses a dilemma that is typical for the construction of all models (both conceptual and material): the trade-off between realism and simplicity. The other chapter on professional wargames, covering the decade between 1870 and 1880, has

5 Welzer, *Climate Wars*, 88–100.

a different perspective and argues for the importance of the context of the leadership concept of *mission tactics*. Finally, Herbert Spencer used the twin concepts of *direct and indirect equilibration* (evolution) for his discussion of the initial functionality and subsequent dysfunctionality of war.

In this book I will not try to imitate or emulate colleagues who have given general and sometimes very comprehensive surveys of the ideas of subsequent military thinkers.[6] My focus is on specific concepts that were used to grasp the problems related to the changing and elusive nature of war. Although some of the concepts discussed in this book were developed specifically for military use (e.g. mission tactics), many concepts were already used in a wider context. Understanding the nature of war implies understanding its wider intellectual and historical context. This contextual requirement has guided the structure of most of my chapters. I discuss a military problem or phenomenon, I explain the concepts used by an author to grasp this problem or phenomenon, and then I try to analyse these concepts in their wider context. It is only with this wider contextual analysis that we can regard thinking about war as a 'great and profound' philosophical topic in the history of ideas.

It seemed attractive to start this collection with the big bang of the General Crisis of the Seventeenth Century. I stop in 1900, so that another volume can start with the even bigger bang of the First World War.

6 See for example Paret, *Makers of Modern Strategy*; Gat, *Military Thought*; Coker, *Barbarous Philosophers*; Jäger, *Handbuch Kriegstheorien*; Coetzee, *Philosophers of War*.

CHAPTER 1

The Cat's Grand Strategy

Pieter de la Court on Holland and the Challenges and Prospects of Free-Riding Behaviour during the General Crisis of the Seventeenth Century

1 Introduction

If grand strategy can be defined as 'the art and science of employing all elements of power—diplomacy, economy, military, information—in order to accomplish national or coalition objectives', then the *Interest van Holland* by the Dutch political and economic thinker Pieter de la Court (1618–1685) presented a grand strategy.[1]

The *Interest* was published first in 1662 and republished in 1669. This second edition was used for an English translation that appeared in 1702 and again in 1743, with the title *Political Maxims of the State of Holland* (version used in the present chapter).[2] The Dutch Republic in the 1660s saw an intensification of ideological warfare between adherents of the House of Orange and the Reformed Church on one side and the so-called 'True Freedom' faction of wealthy regents on the other side. The latter faction was led by Grand Pensionary Johan de Witt (1625–1672). He was a friend of De la Court and had been involved in the composition of the *Maxims*.[3] The work lambasted the adherents of the Prince of Orange and attacked Reformed preachers as intolerant power-seekers. So virulent was the work, that in spite of De Witt's involvement, and even before William III took over control as the most powerful Dutch politician in 1672, the 1669-version of the book was banned and its author debarred from the Lord's supper.[4] De la Court was an isolated figure who opposed not only the House

1 Leonhard, 'From Operational Art', 221. For a similarly wide definition see Baylis, 'Introduction: Strategy', 4–6; Freedman, *Strategy*, xi.

2 Pieter de la Court, *Interest van Holland* (Amsterdam: J.C. vander Gracht, 1662); Pieter de la Court *Aanwysing der heilsame politike gronden en maximen van de republike van Holland en West-Vriesland* (Leiden: Hakkens, 1669); translations: Pieter de la Court, *The True Interest and Political Maxims of the Republick of Holland and West-Friesland* (London: no publisher, 1702); Pieter de la Court, *Political Maxims of the State of Holland* (London: J. Nourse, 1743).

3 Wildenberg, *Johan & Pieter de la Court*, 19–22.

4 Israel, *Dutch Republic*, 758–66.

 | DOI:10.1163/9789004536678_003

of Orange but also the oligarchic and aristocratic tendencies of the regents who were losing contact with their entrepreneurial origins.[5] He tried to show that Holland's liberty and prosperity was served better by the representatives of its entrepreneurial class than by its princes and stadtholders. He used this vision on the past and present to develop his plans for safeguarding Holland's interests in the future.

Although the political portent of the *Maxims* is clear, the secondary literature is divided about its precise nature.[6] The philosophical, economic, political, military, and diplomatic aspects in De la Court's work have all been noted. An appreciation of both the descriptive and the prescriptive elements of his approach is not lacking either. I shall argue that these aspects and elements were part of a remarkably consistent grand strategy for Holland in relation to its Dutch allies and the European powers. I shall present an outline of this strategy that was built around the accomplishment and defence of commercial goals; sketch a historical context that takes into account the general historical shift from tribute-taking agrarian societies towards commercial wealth-generating polities; consider the violent contemporary military and ideological background against which De la Court's strategy stands out; try to reach a better understanding of his strategy through the use of three basic concepts taken from game theory (prisoner's dilemma, assurance game, and free-riding); and stress the distinctive character of De la Court's work, by comparing the practical and strategic use of these game theoretic concepts in the *Maxims* with the function of the same concepts in the philosophical contract theories of Hobbes and Spinoza.

2 Commerce and Compromise

Pieter de la Court was the son of a rich Walloon cloth merchant. He studied at Leyden university and became a cloth merchant himself. He wrote extensively on politics and political philosophy, often in close cooperation with his brother Johan (1622–1660).[7] When De la Court writes about the interest of Holland, he is emphatically not using 'Holland' as *pars pro toto* for the Dutch Republic of the United Provinces. The Union of Utrecht (1579) and subsequent treaties

5 Weststeijn, *Commercial Republicanism*, 26–37.

6 Cf. Kossmann, 'The Course', 62; Nyden-Bullock, 'Radical Cartesian Politics', 50; Malcolm, 'Hobbes and Spinoza', 548 and 549; Soll, 'Accounting for Government', 227; Weststeijn, *Commercial Republicanism*, 346.

7 Weststeijn, *Commercial Republicanism*, 26–37.

had inaugurated a loose confederation of the northern provinces of the Low Countries against the Spanish rule of King Philip II, in which provisions had been made for a common army and navy and the joint defence of fortified towns and cities. Political sovereignty rested with the individual provinces and the main towns within these provinces. The province of Holland was by far the richest and it was agreed that it alone would furnish 58% of the 'national' budget of the States-General.[8] The perspective of De la Court's *Maxims* was not Dutch but Hollandish. Actually, the specific background of the *Maxims* was even more narrowly local; the *Maxims* were based on an unpublished manuscript on the political interests of De la Court's home town Leiden. Only later was *Het Welvaren van Leiden* (1659, written with his brother Johan) broadened to the provincial level.[9]

De la Court's *ultima ratio* is economic prosperity based on trade, industry, fishing, and navigation.[10] All instruments of power should be used towards the realization of this well-articulated aim. The interconnected way in which he suggests using these instruments makes it possible to apply the term 'grand strategy'.

In the first part of the *Maxims* De la Court observes that Holland's agriculture had never been sufficient to feed itself and that it had turned to fishing and manufactures since the Middle Ages.[11] These activities had stimulated the construction of a large fleet and this in turn had a positive effect on Holland's trade. Thanks to these factors, the Hollanders had become masters of the seas.[12] Moreover, the suffering caused by the Dutch Revolt in the southern Netherlands pushed many wealthy merchants to Holland, which offered freedom of conscience and enterprise.[13] In the first part of the *Maxims* De la Court formulates the following *internal* policy recommendations: fishing, industry, trade and navigation should be favoured; it is 'very useful to that end' that all religions be tolerated; all strangers should be given freedom to dwell in Holland; every inhabitant should be free to choose his own trade and occupation (without interference by governments and guilds); the authorities should be very cautious in levying custom duties; justice should be framed not to the benefit of the authorities but to the interest of the inhabitants, especially those

8 Price, *Holland and the Dutch Republic in the Seventeenth Century*, 221–246.
9 Wildenberg, *Johan & Pieter de la Court*, 20, 130.
10 De la Court, *Maxims*, i.2, 13.
11 De la Court, *Maxims*, i.6, 21.
12 De la Court, *Maxims*, i.13, 48 (margin).
13 De la Court, *Maxims*, i,14–15, 49–59.

of the merchants; and colonies should enable the government to put its poor and jobless citizens to commercial use.[14]

De la Court's views on *foreign* policy, in the second part of the *Maxims*, are also dictated by commercial interests. Matters of war and peace are considered from this perspective and result in a strong preference for the latter over the former.[15] Rather than fighting to the bitter end for decisive victories, Holland should be willing to compromise. Wars deprive Holland of all the trade with the enemy's country, while Holland's ships, goods and debts will be confiscated. Moreover, its trading and fishing ships are dispersed and hence difficult to defend, while little booty is available at sea, since Holland itself is the most important trader. Costly land wars to gain fortified cities by long sieges are not to the benefit of Holland either, although it should take care to fortify its own cities. Holland does not need to wage offensive wars against its European neighbours, since it has natural frontiers that are easily defensible. Nor should it indulge in colonial adventures that put territorial expansion before profitable trading, as had happened during the Republic's brief and ill-fated conquest of Brazil (1630–1654).[16] It might be thought advisable, once the province of Holland is engaged in war, to continue until the enemy has been compelled to a well-grounded peace. However, De la Court remarks that an assured peace is an illusion, and even an uncertain peace is still better than war.[17]

Given the vital contribution of trade to Holland's economy, it is not surprising that De la Court favours a government for Holland that contributes most to its commercial prosperity. He defends a commercial republicanism in which the ruling class consists of 'wise merchants', and in which the preferred republican model is not the militant Roman Republic, but rather the commercial republics of Athens and Carthage.[18] He points out that the commercial city of Carthage kept its maritime trade empire as long as it had a free government 'under two *Suffetes*, or yearly burgomasters'—which evokes a clear parallel with the Dutch Republic.[19] De la Court's republicanism is underpinned with

14 De la Court, *Maxims*, ii.15, 307–8.

15 De la Court, *Maxims*, ii.2, 195 (title): 'Above all things war, and chiefly by sea, is most prejudicial, and peace very beneficial for Holland'.

16 De la Court, *Maxims*, ii.15, 307–8.

17 De la Court, *Maxims*, ii.4, 206: 'war is much worse than an uncertain peace, and among all pernicious things, except the intollerable slavery of being governed by the will of a single person, nothing is more mischievous than a war'.

18 Weststeijn, *Commercial Republicanism*, 348.

19 De la Court, *Maxims*, iii.3, 365–4. For another comparison between Rome and Carthage, by Montesquieu, see below §3.2.

a rigorously self-interested theory of motivation.[20] Human behaviour is dictated by short-term passions, both on an individual level and on the collective level. On both levels self-interest is paramount: 'in all societies or assemblies of men, self is always preferred'.[21] Governments will always seek their own interests. Consequently, the interests of the people are best served by a form of government in which the interest of rulers and ruled coincide.[22] In such states, the elite will serve the commonwealth by serving itself. Since the 'interest' of Holland consists of its trade, fishery, and manufactures, it should be a republic ruled by members of its entrepreneurial class.[23] These rulers will be motivated to keep peace as long as possible while at the same time vigorously defending Holland's trade routes.

De la Court's believes that the power and wealth of rulers is inextricably linked with tax revenues and hence with a large population. In addition, he points out that the wealth of this population depends on trade and export of its own products. Finally, he assumes that the total volume of international trade is constant, so that an increase in the share of one nation will always result in a decrease of the share of other nations. These assumptions are typical of a mercantilist economic policy—although De la Court's mercantilism is not aimed exclusively at the protection of exports at the cost of imports, but rather at the protection of the transit trade of Holland's staple markets; hence his dislike of custom duties.[24]

De la Court's preference for a commercial republic headed by entrepreneurs, was mirrored by a strong anti-monarchism. Princes have an interest in manageable and servile subjects. Hence, they will tend to curb the greatness and power of their cities by excessive taxing, so that these will not be able to raise armies against their monarch. Given this observation, the worst thing that could happen to Holland would be rule by a monarch, i.e. a member of the House of Orange.[25] According to De la Court, the princes of Orange had neglected the defence of trade and navigation and exhausted the treasure of Holland with attempts at territorial expansion.[26] Holland needs peace,

20 Weststeijn, 'Passion of self-love', 79–80; Weststeijn, *Commercial Republicanism*, 71–87.

21 De la Court, *Maxims*, i.1, 2.

22 De la Court, *Maxims*, i.1, 1: 'the true interest of all countries consists in the joint welfare of the governors and governed'.

23 De la Court, *Maxims*, iii.1, 312.

24 See also Th. van Tijn, 'Pieter de la Court' xi–xiv.

25 De la Court, *Maxims*, i.1, 6: 'And therefore I conclude, that the inhabitants of *Holland*, whether rulers or subjects, can receive no greater mischief in their polity, than to be governed by a monarch, or supreme lord'.

26 De la Court, *Maxims*, ii.1, 132–195; De la Court, *Maxims*, iii.2, 336–355.

but futile glory and dynastic interests make monarchs more inclined to military adventures than republics.[27] De la Court's polemic comparison between regimes dominated by entrepreneurs and those led by princes also influenced his defence of compromise over intransigence. An aptitude for making compromises is closely linked with a facility for calculating the probability of different scenarios and decision-making based on these calculations. De la Court suggests that merchants by the very nature of their profession excel in this activity, and that a successful state has much to benefit from a calculating government—hence the theme of the 'wise merchant'. De la Court devotes *Maxims* II.3 to several instances where a *prima facie* case could be made for an offensive war, only to reject this possibility on closer deliberation and calculation. For instance, foreign governments who impose inordinate tolls on goods that Holland desires to export, seem a direct threat to its trade interests. For a nation living by trade this might seem a plausible *casus belli*. Even in this case, however, Holland should not declare war. It should retaliate by charging for the commodities from the foreign power in return. This will cause a drop in the exports of the foreign nation. Moreover, that nation's own new impositions will curb its imports as well. The net effect will be a decline in the foreign nation's general trade volume, and the resulting decline in profits will prompt it to lift its impositions, without any necessity of war.[28]

In general, Holland should never start a war, rather 'we ... ought always to wait till others make war upon us, directly and indeed'.[29] Peace is so important that Holland should not hesitate to comply with blackmail: 'they that buy their peace do best', for 'he who will not bestow a stiver to keep peace, must have his sword always drawn' and will eventually exhaust himself.[30] Holland should not let its strategy be influenced by the obsolete code of honour of 'the idle gentry, soldiers of fortune, and the sottish rable [*sic*]', i.e. the princes of Orange and their adherents.[31] As long as Holland can peacefully enjoy the benefits of being Europe's foremost trading nation, neither territorial dynastic wars nor maritime mercantilist wars are an attractive option.[32] But De la Court's preference for a defensive strategy did not rule out aggressive action as soon as Holland's

27 De la Court, *Maxims*, iii.2, 352.

28 De la Court, *Maxims*, ii.4, 203–4.

29 De la Court, *Maxims*, ii.4, 207.

30 De la Court, *Maxims*, ii.4, 211–212; De la Court, *Maxims* ii.4, 211: 'For peace-sake we ought to yield some-what'. See also De la Court, *Maxims*, iii.2, 352; Velema, 'That a Republic is Better', i, 9–25; Jonathan Israel, 'The Intellectual Origins', 7–36.

31 De la Court, *Maxims*, ii.4, 215.

32 De la Court, *Maxims*, ii.1, 133.

trade-routes were directly threatened. The first and longest chapter of *Maxims* II, devoted to foreign relations, is concerned with the imperative of maintaining and defending free maritime navigation 'against all pirates and enemies'.[33] And although De la Court would prefer to keep England at bay with 'good words', in a passage of the *Maxims* that was absent from the earlier 1662 version of the *Interest,* and which must have been inspired by the Dutch successes of the Second Anglo-Dutch War (1665–1667), he remarks contentedly that if war with England cannot be averted, 'It is certain, that *England, Scotland,* and *Ireland,* having in all parts a deep and bold coast, their cities, towns, and villages in the country being weak, or without wall and fortifications, they may in all places be attacked, and our men may be landed under the shelter of our cannon, and so plunder and burn those places'.[34]

3 Historical Context: Change and Crisis

What was the wider historical context of De la Court's mercantilist, pragmatic, and militarily defensive grand strategy? From about 1200 onwards Europe saw a gradual increase in the weight of commerce in addition to agriculture as a source of economic wealth, an increase that accelerated with the onset of modernity. In the early modern era Europe became the greatest hub in a global exchange network. The relative volume, variety, and intensity of commercial exchange in Europe was greater than anywhere else on the Eurasian continent. In Europe and elsewhere, elites had consisted predominantly of officials and warriors who had used various forms of coercion to extract surplus wealth from their farmers. Since technological innovation and economic growth in agrarian societies were slow and limited, predation was a relatively efficient form of wealth extraction in the short and medium term. Economic expansion was achieved by territorial conquest rather than by internal development. Gradually, as entrepreneurial wealth generation and its associated higher growth rates and increased technological development gained momentum, an

33 De la Court, *Maxims*, ii.1, 132 (title).

34 De la Court, *Maxims*, ii.10, 242; De la Court, *Maxims*, ii.9, 240. The English translation agrees with the 1669 Dutch version, the *Aanwysing*, ii.9, 296. While De la Court assumes in the *Aanwysing* that the Dutch can do more harm to the English than the other way round, he still subscribed to a more pessimistic view in the earlier 1662 *Interest*, 175: 'dat een oorlog aan wederzijden zeer schadelik is; zo nochtans dat Engeland veel beuiten op ons, wy weinig op de Engelsen konnen doen; en dat zy door ons wel zeer beschadigd, maar wy door haar genoegzaam geruineerd konnen werden'. On Anglo-Dutch rivalry see also Jardine, *Going Dutch*, 319–48.

escape from the earlier predation trap became possible. A greater dependence on commercial revenues and a policy of mercantilist trade protection jointly supplanted the previously destructive cycle. The status of merchants increased, and in some polities, including Venice and Holland, merchants became the chief magistrates, adding still further to a self-reinforcing cycle that was no longer parasitic on existing wealth but rather stimulated the accelerated generation of new wealth.[35]

A distinction between an aristocracy that was parasitic and bellicose, and a new entrepreneurial elite that was productive and peaceful, may amount to oversimplification of extremely complex social structures and processes, but both classes had their acute contemporary portraitists. Niccolò Machiavelli gave a sharp portrait of the prince who 'has no other object and no other interest and takes as his profession nothing else than war and its laws and discipline'.[36] A century and a half later, De la Court combined his scathing attack on a parasitic and aggressive aristocracy with a vigorous defence of the merchant-ruler. For him this polemic was inextricably linked with an institutional preference for a republic over a monarchy and a political preference for the 'True Freedom' faction over the princes of Orange. De la Court's *Maxims* can be read as the expression of a strategy that reflected a highly significant historical process in which agricultural tribute-taking was gradually complemented by commercial wealth-generation.

Shifting our perspective from De la Court's views on Holland's socio-economic development to its strategy in relation to the wider world, the most relevant context is formed by what has been called the General Crisis of the Seventeenth Century. De la Court's credo of defensive prosperity must be understood against the background of an age that was in deep and violent crisis.[37] The world in the mid-seventeenth century was rocked by more cases of state breakdown than any previous or subsequent age. Ming China collapsed in the 1640s, the Polish-Lithuanian Commonwealth disintegrated, much of the Spanish monarchy seceded and all the territories of the Stuart monarchy rebelled.[38] The number of popular revolts in Europe, Russia, China, and Japan also rose sharply. Finally, the number of wars, which was high throughout the seventeenth century, peaked in the 1640s.[39] The mid-seventeenth-century not

35 North, *The Rise of the Western World*, 91–156; Mokyr, *The Lever of Riches*, 78–9; Christian, *Maps of time*, 316–24, 393–401; Gat, *War*, 506.

36 Machiavelli, *The Prince*, i, 55.

37 Parker, *Global Crisis*. For a critical historiographical discussion of the concept, see De Vries, 'The Economic Crisis', 151–5.

38 Parker, 'Crisis and Catastrophe', 1052–79.

39 Parker, 'States Make War', 15.

only saw a rise in the *frequency* of interstate wars involving the great European powers, but also in the *duration* of these wars, their *extent* (measured in the number of participating powers), their *magnitude* (measured in nation-years, i.e. the number of nations multiplied with the number of years fought by each nation), their *severity* (measured in battle deaths) and their *concentration* (number of battle deaths divided by nation-years); see Figure 1.[40] Similar trends can be discerned for the Ottoman, Mughal, and Chinese empires.[41]

The General Crisis was caused by a confluence of various factors. Mean temperatures in the entire seventeenth century were significantly lower than before and after, prompting leading historians and climatologists to call the period the 'Little Ice Age'. The middle of this century was especially cool. A rare series of major volcanic explosions between 1638 and 1643 threw tons of ashes into the stratosphere. Moreover, solar energy received on earth diminished, due to an almost complete disappearance of sunspots between 1643 and 1715. These phenomena coincided with an increased number of episodes of El Niño, which produced prolonged drought in some areas. While two El Niños every decade is the recent average, seven occurred between 1640 and 1661. The impact of these climatic upheavals on agricultural production was especially dramatic at higher latitudes.[42] Solar cooling had a stronger effect at higher latitudes than at the equator. At higher latitudes increased snow and ice cover reflected the sun's rays back into space, thus producing a self-reinforcing effect. At these latitudes a relatively modest drop in mean temperatures caused a significant increase in the frequency of cataclysmic floods, storms, prolonged droughts, and abnormally cold spells.[43] David Zhang has argued for a causal relationship between seventeenth-century decline in temperature, falling agricultural production, rising wheat prices, and an increase in the frequency in wars, both in Europe and in China. Food scarcity led either directly to resource-oriented wars, or caused social conflicts and instability which eventually triggered wars as well.[44]

Climate change reinforced the interplay between two factors that further exacerbated the disastrous toll of war in the middle of the seventeenth century. Firstly, the same economic logic that gradually marginalized parasitic warrior elites in favour of entrepreneurial elites, also allowed European powers to allocate greater resources to their armies and navies. Early modern Europe

40 Levy, *War*, 112–36.

41 Parker, 'Crisis and Catastrophe', 1056; Brecke, 'Violent Conflicts'.

42 Parker, 'States make War', 17–20.

43 Parker, 'States make War', 17–20.

44 Zhang, 'Global climate change', 19217–8; see also Parker, *Global Crisis*.

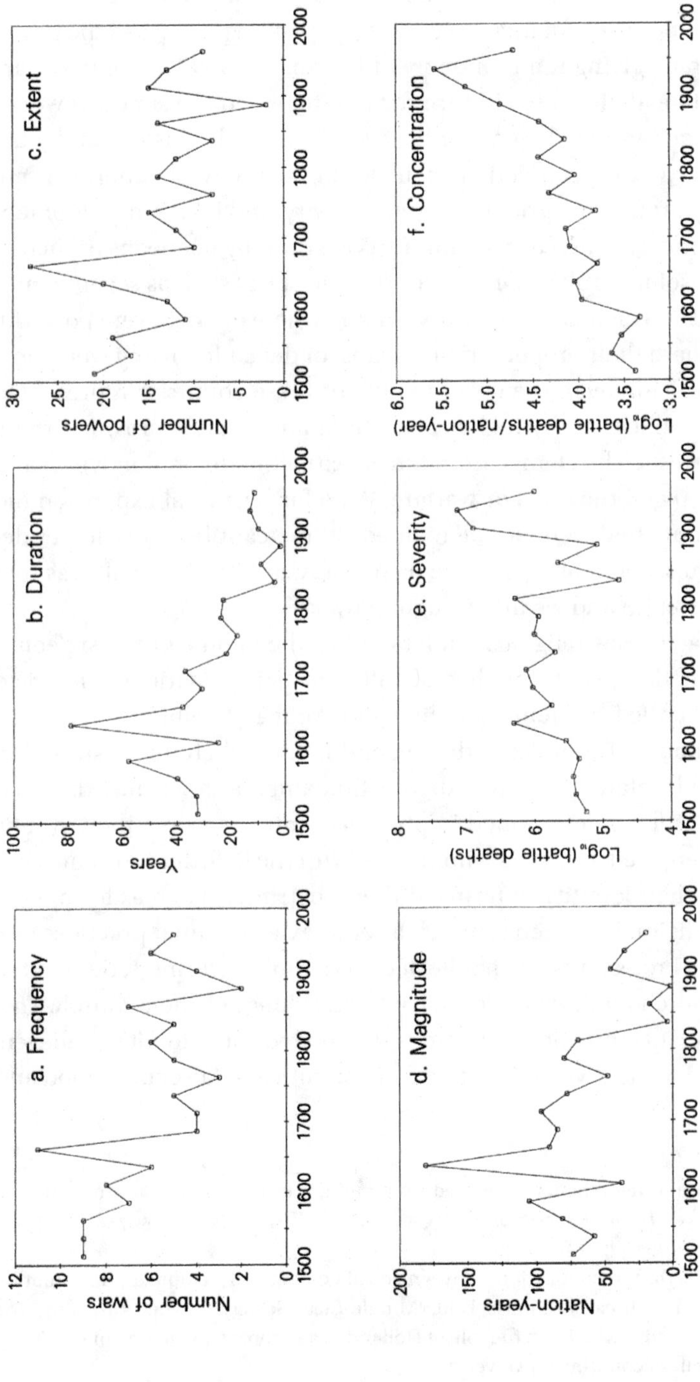

FIGURE 1 Wars between the european powers, in twenty-five-year periods, 1500–2000
SOURCE: JACK S. LEVY, *WAR IN THE MODERN GREAT POWER SYSTEM, 1495–1975* (LEXINGTON, KENTUCKY: UNIVERSITY PRESS OF KENTUCKY, 1983), 133

experienced a military development in which firearms transformed both field and siege warfare. Moreover, armies and navies grew in size and became more permanent.[45] Money used for increased military spending increased the power of the central state, giving it a monopoly of legitimate force, making warfare all the more a state affair, which in turn further increased the state's power of taxation and command.[46] The economic side of this cycle was driven by the income from extra-European activities, including the flow of American bullion, by a steady growth in agricultural productivity, and by the development of deficit financing. The increase in military power was by no means limited to Europe's new absolutist rulers. On the contrary, in states such as seventeenth-century Holland and eighteenth-century Britain, where the rich were powerful enough to safeguard their property rights, loans for deficit financing were most readily available. Moreover, the governments of these countries represented elites that were willing to pay higher taxes to finance wars—provided these served their interests. The change from forced extraction to productive wealth creation also changed the aims of warfare. Wars for territorial expansion did not disappear, but they were supplemented by mercantilist wars for trade-posts, trade-routes, and markets. In the words of Azar Gat, the result was that 'power brought wealth and wealth brought power'.[47]

The second factor was religious intolerance. In the middle of the sixteenth century, Western European churches of different denominations embarked on a process described by Heinz Schilling and Wolfgang Reinhard as confessionalism. The Roman Catholic, Lutheran, and Reformed churches started to issue precise and detailed formulations of fundamental doctrines that had been left remarkably undetermined in previous centuries, when the Catholic church had still enjoyed a religious monopoly. With the Reformation, churches began to define their identity in terms of their doctrines. Whereas for medieval Christians religion had been as much a set of external ritual practices as a set of beliefs, the Protestants, soon followed by Catholic reformers, demanded that ordinary church members internalize the teachings of their church. This rise of mutually antagonistic confessions stimulated dichotomies. Believers tended to see the world divided into two camps: truth and falsehood, good and

45 Whether these developments amounted to a 'military revolution' remains a contested issue; see Parker, *The Military Revolution*, 155–176; Gat, *War*, 456–71, 476, 507–8.

46 Gat, *War*, 443–511.

47 Gat, *War*, 508. On the connection between social-economic development and military succes in the Dutch case, see also Feld, 'Middle-Class Society', 419–42; Downing, *The Military Revolution*, 212–238. On the role of Holland as a European paradigm for economic reform see Soll, 'Accounting for Government', 217.

evil, Christ and Antichrist. Confessionalization not only sharpened differences between denominations, it also stimulated greater uniformity within each creed.[48]

The mutually reinforcing effects of military state power and confessionalism had explosive results that brought Europe's rulers both perils and opportunities. A difference in religion could alienate subjects from their prince, but a shared religion could bolster a ruler's authority. Rulers tried to impose religious uniformity on their territories. Religious enemies tended to become political enemies and vice versa. The result was a fusion of religious and political identity, of piety and patriotism.[49] Thus warfare, fed by religious intransigence and an ever more efficient state apparatus, against a background of climate-induced misery, was intensive throughout the seventeenth century, with a peak in the middle of the century. And this is precisely the period in which De la Court formulated a grand strategy that was built around trade, tolerance, and a defensive foreign and military policy that was so flexible that it would even accept blackmail—all for the sake of Holland's wealth.

4 Game Theory

If De la Court used his *Maxims* to develop a grand strategy for Holland, then it might be useful to analyse his work in terms of a discipline that studies strategy, i.e. game theory. Game theory tries to capture behaviour in strategic situations, in which the choice of one player (e.g. person, group, or nation) depends on the choice of one or more other players, and in which the choice of each player is motivated by a self-interested rationality.[50] Game theory has developed into an extremely complex and sophisticated discipline, and I will limit myself to an elementary description of three basic 'games' that will be used for a discussion of De la Court and for a comparison with Hobbes and Spinoza.

The first game is a simple prisoner's dilemma.[51] The parties make simultaneous decisions, and each party can choose between cooperation or non-cooperation with the other party. For each option different payoffs can be specified. It is possible to specify a payoff-matrix in which for each player

48 Reinhard, *Die katholische Konfessionalisierung*, 1–49, 419–52; see also Kaplan, *Divided by Faith*, 28–47.

49 Kaplan, *Divided by Faith*, 99–103.

50 See Morrow, *Game Theory*; De Bruin, 'Game Theory', 197–208; Rasmusen, *Games and Information*.

51 Rasmusen, *Games and Information*, 19–21.

TABLE 1 Prisoner's dilemma

		Nation B	
		Peace	War
Nation A	Peace	(2) (2)	(4) (1)
	War	(1) (4)	(3) (3)

EXPLANATION: PRIORITIES OF NATION A ARE GIVEN FIRST, FOLLOWED BY THOSE OF NATION B; INCREASING NUMBERS INDICATE DECLINING PRIORITIES

non-cooperation is always more rational than cooperation, even though the rewards of joint cooperation are greater than the rewards of joint non-cooperation. An instance would be two nations that have the choice between waging war or not waging war on each other, with the following outcomes for each nation in *declining* order of priority: (1) wage war on a peaceful neighbour; (2) live in mutual peace; (3) live in mutual war; (4) refrain from war and suffer invasion (see Table 1).

The second game is a so-called 'assurance game'. In this game nation A and B again have both the option to cooperate or not to cooperate, e.g. to keep peace or to attack. Yet the dividends are different. For each player, dividends are highest when both players cooperate. In this case, non-cooperation is no longer the only rational strategy. Cooperation is an option as well. It is even the preferred option, but it is also a precarious option that depends on the rationality of both players and on a trust of each player in the rationality of the other. If one player acts irrationally or if one player has a justifiable fear for an irrational act of the other, then cooperation will break down and war rather than peace will become the dominant strategy (see Table 2).[52] While the prisoner's dilemma is a one-shot game, assurance games are typically played in a sequence of repeated games. This allows trust to develop, although the possibility of a collapse of trust remains an ever-present possibility.

The rules of the third game are determined by joint (or collective) action problems. In the words of Seumas Miller, 'A situation involves joint action if: there is more than one agent; each agent is performing (at least) one action; [and] each agent's action is dependent on the actions of the other agents'.[53]

52 Hampton, *Hobbes*, 65–66; Moehler, 'Hobbes' State of Nature', 307.

53 Miller, 'Joint Action', 275.

TABLE 2 Assurance game

		Nation B	
		Peace	War
Nation A	Peace	(1) (1)	(4) (2)
	War	(2) (4)	(3) (3)

In many contexts, all the individual members of a group can benefit from the efforts of the group, when these efforts result in the production of collective goods, e.g. we can all benefit from the roads that are constructed and maintained by our taxes. Each member of the group will benefit even more, however, if he can continue to use the collective good without making his own individual effort. This is called free-riding.[54] Free-riding becomes a more tempting option when the group consists of many individuals. The larger the number of individuals, the easier one individual can stop contributing without seriously affecting the total amount of the collective good produced. Moreover, increasing numbers will often result in increasing coordination problems between the individuals, thus adding another stimulus for the occurrence of free-riding behaviour.[55] The most salient feature of joint action problems is that the priorities of the group are not necessarily the same as those of single persons. The collective good can be considered so vital that the group will continue to deliver in spite of individual free-riders (see Table 3). But no collective good will be produced if everyone starts to free-ride; so the stability of this game is very much dependent on the efficiency of a mechanism by which the number of free-riders is held in check.

54 See Hampton, 'Free-rider problems', 245–73; Tuomela, 'On the Structural Aspects', 165–202. The concept of free-riding that I ascribe to De la Court is part of a modern game theory that was given its first general mathematical formulation by John von Neumann and Oskar Morgenstern in 1944 (Neumann, *Theory of Games*). However, the concept, if not the term, of *free-riding* was possibly already what Plato had in mind in the *Laches* and the *Symposium* when he presents Socrates recalling an episode from the Battle of Delium (424 BC; See Ross, 'Game Theory'). And Thomas Hobbes and Benedictus de Spinoza certainly tried to solve the problems caused by free-riding in their political philosophy, see below §1.7.

55 This point is well made in Hume's classic formulation of the problem, see Hume, *Treatise*, III.ii.7, 538; see also Baldwin, 'Hume's Knave', 277–96.

TABLE 3 Joint action and free-riding

		Group	
		Deliver	Defect
Single person	Deliver	(2) (1)	(4) (3)
	Defect	(1) (2)	(3) (4)

5 Holland and Europe: the Opportunities of Free-Riding

Holland and the other provinces of the Dutch Republic reached the zenith of their Golden Age between the Peace of Westphalia in 1648 and the catastrophic military reverses of 1672, after a joint attack by France, England, the Prince-bishop of Munster, and the Elector of Cologne. Since Holland's apogee coincided with the General Crisis of Europe, the question arises whether this peak was not merely reached *in spite* of the Crisis, but even *thanks* to the Crisis.[56] This could imply a form of free-riding (see Table 3). De la Court had indeed a remarkably keen eye for various forms of free-riding. For instance, while the protection of Holland's maritime commerce dictates a complete eradication of piracy in the North Sea, protection against the same threat in the Mediterranean demands a different course. In this region, English, Spanish, Italians, and competitors from the Levant would profit too much from an effort by Holland to eradicate piracy, i.e. they would free-ride on the efforts by Holland. So, in order to prevent free-riding by the merchants of other nations, 'who by that means, and other advantages, might easily deprive us of our traffic and freight ships', De la Court suggests a different strategy: Holland's warships should convoy its traders, and the captains of these warships should not be allowed to protect foreign merchants, 'so that if we should leave this thorn of the *Turkish* pirates in their sides, they will be sufficiently distress'd both in that and all their other trade, whilst we by those ordinary convoy ships of war, may wholly engross all the *European* traffick and navigation to *Holland*'.[57]

In another example De la Court remarks that alliances are best kept by parties that are subject to the sanctions of a more powerful party, i.e. parties

56 Cf. Schöffer, 'Holland's Golden Age', 78–109.

57 De la Court, *Maxims*, ii.1, 135.

that can punish free-riding.[58] But where compliance by one party has to be enforced, and where more than one candidate for the role of enforcer is available, each of the potential enforcers will try to relegate the task of enforcing to one of the other parties. Potential enforcers will prefer to free-ride on their duties: 'no body would willingly be the compeller, but every one would ride on the forehorse', which sums up another free-riding problem.[59] These are fairly unproblematic examples of (attempts to prevent) free-rider behaviour. De la Court discusses other cases that are conceptually more interesting.

Since economic prosperity is the key to De la Court's grand strategy for Holland, and since war with any nation is detrimental to Holland's trade with that nation, it is not surprising that he advocates peace with all nations. At the same time he is not opposed to wars between Holland's neighbours. He even stresses that Holland owes its prosperity in large part to the wars between its neighbours that marked the seventeenth century. The combating parties stimulated Holland's military manufactures, while their refugees provided the Dutch with new industrious citizens.[60] Since Holland should profit from the wars amongst its neighbours while remaining neutral itself, it should take care not to become entangled in alliances, which anyway tend to last only as long as they are useful to both parties.[61] Hence no alliance with a stronger power should be made, unless the stronger power first performs its contract.[62] In general, the rest of the world should be left to its own devices while Holland sits still behind strong frontiers in impregnable cities and devotes its energy to further improving its army and its fleet.[63]

De la Court's defensive strategy forms the background for his comment on Machiavelli's well-known dictum that rulers need to be able to put on the skin of a lion as well as that of a fox.[64] In full agreement with his republicanism noted above, De la Court points out that this may apply to princes but not

58 De la Court, *Maxims*, ii.5, 217.

59 De la Court, *Maxims*, ii.5, 217.

60 De la Court, *Maxims*, ii.1, 194: 'those lasting wars, and terrible devastations of *Germany*, and many other adjecent countries, supported and supplied our cities with manufactures, merchants and mechanicks; who finding here the states manner of government not quite overthrown, have under those remains of publick freedom, erected many new manufactures and trades, and have been able to keep up the old imployments and traffick of *Holland*, especially through the dilligence, vigilance, valour, and frugality, which are not only natural to the *Hollanders*, but by the nature of our country is communicated to all foreigners that inhabit among us'.

61 De la Court, *Maxims*, ii.6, 227.

62 De la Court, *Maxims*, ii.5, 221–2; see also De la Court, *Maxims* ii.9, 243.

63 De la Court, *Maxims*, ii.5, 210; see also De la Court, *Maxims* ii.4, 205.

64 Machiavelli, *The Prince*, 65.

to republics. Monarchs are warlike and, like lions, live upon the flesh of their enemies and even that of their own subjects. As soon as the enemies made by these irresponsible actions turn against them, monarchs need the skin of a fox. Republics, on the other hand, because they tend to govern 'with more gentleness, wisdom, and moderation, can naturally rely on more support from their own citizens than monarchies, and therefore stand not in need of such maxims, especially those that subsist by trade'.[65] A trading republic like Holland ought to follow the example of a cat; it stays at home, meddles with no one, hunts alone, preferably takes flight when it is threatened, but fights fiercer than a lion when it cannot avoid combat, '[s]o that by these arts that species enjoy[s] more quiet every where, live[s] longer, [is] more acceptable, and in greater number than lions, tygers, wolves, foxes, bears, or any other beasts of prey, which often perish by their own strength, and are taken where they lie in wait for others'.[66]

De la Court's cat strategy is complicated and can be understood with game-theoretical situations described in the previous section. The chronic warfare itself, from which Holland is supposed to profit, can be described in the grim terms of a prisoner's dilemma (Table 1); warfare results when both parties refuse to cooperate. Moreover, De la Court's strategy seems to imply some form of free-riding (Table 3), but of a very peculiar variety. Holland should not join the wars of its neighbours; rather it should profit from the collective 'good' that these nations produce, i.e. the calamities of war. It is thanks to these disasters that Holland will profit from the influx of new citizens escaping from territories ravaged by war, and its undamaged commercial trading fleet will profit from the damage that belligerent powers have inflicted on each other's ships. This case is stronger than standard cases of free-riding: the free-rider does not profit *in spite* of not contributing, but even *because* he does not participate. Since this is not an instance of profiting in spite of not contributing to a *positive good*, but rather a case of profiting thanks to not contributing to a *negative 'good'*, this could be called 'inverse free-riding'. And since this peculiar form of free-riding is a game-theoretical situation that feeds on its turn on another, preliminary, game-theoretical situation, i.e. the prisoner's dilemma in which belligerent nations are locked (Table 1), it can also be called 'derived free-riding'. De la Court's strategy of derived and inverse free-riding was by its nature precarious and limited. It was precarious because belligerent neighbours who had slowly started to recover from the worst effects of the General Crisis might find it

65 De la Court, *Maxims*, ii.4, 208.

66 De la Court, *Maxims*, ii.4, 209; see also De la Court, *Maxims*, ii.9, 244–5.

attractive to attack the peaceful and wealthy Dutch free-rider, and it was limited because defensive considerations of the Dutch, foreseeing this possibility, forced them to join the costly arms-race of their neighbours as much as if they would have actually participated in their wars. Yet this, according to De la Court, was the key to Holland's success.

6 Holland and the Other United Provinces: the Challenges of Free-Riding

De la Court's preoccupation with free-riding can also be found in his discussion of Holland's relation with the other Dutch provinces. A major theme in the *Maxims* is the author's frustration with the modest contribution made by the other provinces towards the collective military security of the Dutch Republic. While the geographically central province of Holland bears the brunt of the financial burdens, the peripheral provinces who profit most from the protection offered by the Union pay least. De la Court points out that Holland pays even more than it was obliged to by the Union of Utrecht, while its contributions for the upkeep of conquered towns and provinces bring no profit.[67] Moreover, 'almost all the *United Provinces* have continually preyed upon *Holland*, by bringing in very many more provincial charges to the account of the generality'.[68] While Holland pays for almost the entire Dutch fleet and most of its army, the institutions of the Republic give it only a limited say in the admiralties and the States-General that exert naval control.[69] Meanwhile others 'wallow in idleness and gluttony with the wealth of *Holland*'.[70] De la Court complains that this unfortunate state of affairs has been maintained by a coalition between the princes of Orange and the other provinces.[71]

De la Court's analysis of the neglect of the other provinces in providing their share of the budget of the Republic is reminiscent of the free-riding that is characteristic for a joint action problem (Table 3). The Union of Utrecht was meant to provide a collective good to the participants in the treaty. This collective good consisted of military security. The six provinces were obviously loath to deliver their full share. But how well are De la Court's remarks about

67 De la Court, *Maxims*, i.3, 19.

68 De la Court, *Maxims*, ii.11, 257.

69 De la Court, *Maxims*, ii.1, 148.

70 De la Court, *Maxims*, i.23, 88; see also De la Court, *Maxims*, ii.1, 148; De la Court, *Maxims*, ii.10, 249: '*That it is very easy to lie in the ashes with another man's garment, and be warm*'.

71 De la Court, *Maxims*, ii.13, 289.

the six provinces actually captured by the concept of free-riding? Firstly, a total number of merely seven parties seems *small* enough to make serious coordination problems relatively unlikely. However, power within each of the seven provinces rested on their constituent towns (and in some cases also on rural areas), leaving room for free-riding behaviour on the subsidiary level of smaller political units, thus complicating agreement about financial burdens between the provinces and causing real coordination problems. Secondly, a total of six possible free-riders out of a total of seven parties seems so *large* that the production of the collective good threatens to become impossible, in which case free-riding would be a self-defeating exercise. However, the unequal weight of Holland's nominal 58% contribution (even more in practice), for each of the other provinces had the same effect as cooperation with many more equal participants. Hence each of the six provinces could find it attractive to shoulder less than its full share, since for each of these free-riders this would result in only a negligible decrease in the total amount of collective security. Here we have a collective action paradox that is still very relevant for small groups that try to achieve common ends, including military ends. This mechanism has been described by Mancur Olson as the 'exploitation of the great by the small'. In addition to the goal shared by all the players (in this case collective security), the biggest players have a goal that is not shared by the smaller players. The extra goal of the big players is to maintain their dominant position. The small players will assume correctly that the big players will pay for this additional goal anyway, even though the big players will grumble about the free-riding behaviour of the small players.[72] (This mechanism seems at work in the relation between the USA and its NATO partners as well.)

One solution to free-riding behaviour consists in the threat or actual administration of sanctions. Sanctions can be administered, either by an overarching authority that can force the parties involved to share their part in the production of the collective good, or by the participating parties themselves. Since the Republic consisted of sovereign provinces and towns, the first option was not available. De la Court seems to opt for a variant of the second possibility when he considers a termination of the cooperation or a threat of termination. He remarks that Holland should not assist its weak allies more than might agree with its own interest.[73] The Dutch navy should be wrested out of the hands of the other provinces and be brought under the control of those who stand to lose most from damage to Holland's maritime trade, i.e. its merchants.[74] De la

72 Olson, *The Logic*, 3, 27–32, 34–6.
73 De la Court, *Maxims*, ii.11, 265.
74 De la Court, *Maxims*, ii.1, 195.

Court even makes the suggestion that Holland and its close ally Utrecht dig a channel between the Zuiderzee and the Rhine, thus creating a defensive barrier that would be able to protect Holland and its bulwark Utrecht against the threat of foreign invasion, without the need for costly subsidies to the other provinces.[75] This seems an unambiguous call to end cooperation with the other provinces, but another aim of this remarkable proposal may have been to pressurize the provinces at the periphery into finally contributing what Holland considered to be their due share.

So, while in the case of Holland's relations with the European powers De la Court's cat-strategy implied an *actual* opt-out of European warfare, the *threat* to opt out of the Union of Utrecht formed the centrepiece of a similar strategy in relation to the six other provinces.

7 Philosophical Context: Hobbes and Spinoza

Various sources have been suggested for De la Court's political and military ideas, with Machiavelli's realism, Descartes's psychology based on his theory of the passions, Hobbes's notions of self-interest and absolutism as the main contenders.[76] De la Court himself, in his turn, is often mentioned as a source of influence on Spinoza.[77] In *The Cambridge History of Political Thought 1450–1700* Noel Malcolm uses De la Court as a 'bridge' between Hobbes and Spinoza.[78] Since I have analysed De la Court's grand strategy in terms of game-theoretical challenges and opportunities, with special attention to free-riding, it may be useful to compare his views on these topics with the ideas of Hobbes and Spinoza.

Game-theoretical problems play a central role in Hobbes's political works.[79] His well-known contract theory can be regarded as a solution to these problems. In the state of nature, constant mutual fear makes it difficult to establish cooperation. In *The Citizen* (the 1651 translation of *De Cive*, 1642), Hobbes explains that fear is partly caused by men's 'mutuall will of hurting' and partly

75 De la Court, *Maxims*, ii.14, 291 ff.

76 See Van Heck, 'In het spoor', 277–318; Kossmann, 'The Course', 65–6; Nyden-Bullock, 'Radical Cartesian Politics', 48–51; Malcolm, 'Hobbes and Spinoza', 474; Burke, 'Tacitism', 484; Weststeijn, *Commercial Republicanism*, 83–5.

77 See Malcolm, 'Hobbes and Spinoza', 553; Kossmann, 'The Course', 75.

78 Malcolm, 'Hobbes and Spinoza', 547–50; cf. Weststeijn, *Commercial Republicanism*, 18–19, 57, 142–57, 356.

79 See Kavka, 'Hobbes's War', 291–310; Hampton, *Hobbes*, 58–79; Moehler, 'Hobbes', 297–326.

by their natural equality. A roughly equal physical and mental strength allows each individual to hurt any other individual.[80] The result, as Hobbes puts it famously in the *Leviathan* (1651), is a war 'of every man, against every man'.[81] This situation is obviously captured by the prisoner's dilemma in Table 1. According to Hobbes, the way out of this predicament is a social contract. Once this contract has established a civil society, including mutual cooperation that aims at the production of public goods, free-riding behaviour (see Table 3) should be restrained by fear of punishment.[82] This fear should be provided by a central authority that has the power to curb free-riders by the application of sanctions, 'whereby particular men may be ruled through fear of punishment'.[83]

Since the whole point of Hobbes's social contract seems to be the establishment of an internal peace that allows cooperation, it is remarkable that he explicitly appreciates the possibilities for cooperation prior to conclusion of the contract. In the *Leviathan* he mentions the fool who (in the state of nature) 'hath sayd in his heart, there is no such thing as Justice'.[84] The fool reasons that since every human being must fight for its own preservation, there is no reason to keep promises and contracts. Against the fool Hobbes argues that it can be rational to keep contracts: 'But either where one of the parties has performed already; or where there is a Power to make him performe; there is the question whether it be against reason, that is, against the benefit of the other to performe, or not. And I say it is not against reason'.[85] The second possibility, concerning the presence of a power that enforces parties to perform, clearly assumes an ordered society, but the first possibility can also occur in the pre-contractual state of nature. In the state of nature each individual has good reason to fear all his fellow-creatures, so each man is very much in need of allies or 'Confederates' that help him in his quest for survival. Hence it may be more rational for both parties to cooperate rather than not cooperate. Yet Hobbes's condition for this cooperation is that the other party performs first. So how can cooperation ever start? Actually, if both parties are convinced of each other's rationality, it will be rational for both to perform first, so that the requirement will become meaningless.[86] We have seen that this assumption

80 Hobbes, *The Citizen*, i, 113.

81 Hobbes, *Leviathan*, I.xiii, vol. 2, 192.

82 Hobbes, *The Citizen*, v, 167.

83 Hobbes, *The Citizen*, v, 169.

84 Hobbes, *Leviathan* I.xv, vol. 2, 222. Psalm 14.1: 'The fool hath said in his heart, There is no God'.

85 Hobbes, *Leviathan*, I.xv, vol. 2, 224.

86 For a fuller explanation of the paradox and its solution see Hampton, *Hobbes*, 65.

of mutual rationality is typical of the assurance game (Table 2), which seems indeed a good model for the kind of cooperation suggested by Hobbes. If Hobbes had considered this pre-contractual model a fully viable alternative, then he would have been obliged to explain the need for any subsequent social contract and the associated grim authority of an absolutist state. But Hobbes's theory of the passions implies that irrational behaviour and fear of irrational behaviour makes a collapse of the assurance game into a prisoner's dilemma a constant possibility; hence his preference for the alternative of a civil society under an absolute sovereign.[87]

Many game-theoretical problems in Hobbes's discussion of the transition from a state of nature to a contractual society can also be found in the works of Spinoza, especially in his *Tractatus theologico-politicus* (1670), although this aspect of his work has received less attention than Hobbes.[88] In the state of nature, the natural right of each individual man is determined by short-term passions rather than sound reason. Yet it is clear that a life in accordance with laws and the dictates of reason would be preferable. From this principle follows the necessity of a social contract or 'compact'. If every man were capable of having his actions guided by the long-term utility of the compact there would be no problem. Unfortunately, after the conclusion of the compact, men are still led astray by passions rather than reason. Hence promises made when the compact is concluded amount to little, 'unless something else is added'.[89] What is this something? While in the state of nature the natural rights of individuals are only limited by their power, part (but not all) of these rights comes to rest in the hands of the sovereign after conclusion of the contract. Hence the sovereign is in a position to punish free-riders. Although Spinoza writes that the individuals have only transferred 'a part' of their original rights, the power of the sovereign can nevertheless 'compel men by force, or restrain them by threats of the universally feared punishment of death'; and the subjects 'are obliged to fulfil the commands of the sovereign power, however absurd these may be'.[90]

Although Spinoza's discussion of the transition from natural to civil state in the *Tractatus theologico-politicus* shares important elements with Hobbes's account, there are at least two differences. Firstly, unlike Hobbes, Spinoza

87 Hampton, *Hobbes*, 80–9; see also Moehler, 'Hobbes', 300 n. 5, 306.

88 See Rosenthal, 'Two Collective Action Problems', 389–409. On the presence of a contract theory in the *Tractatus theologico-politicus* and its absence in the *Political Treatise*, see Matheron, 'Le problème de l'évolution', 258–70.

89 Spinoza, *Theological-Political Treatise*, ii, 286.

90 Spinoza, *Theological-Political Treatise*, ii, 287–8.

pays scant attention to the possibilities of cooperation in the state of nature, i.e. to the prospects offered by the assurance game. For Spinoza this is not a serious possibility. Hence, he jumps straight from the problems of mutually destructive behaviour in the state of nature to the initiation of a compound that results in an ordered civil society, with a powerful sovereign as the means of upholding stable cooperation and curbing free-riders.[91] Secondly, although Spinoza is more inclined than Hobbes to put his faith in civil society, he has at the same time more reservations than Hobbes about the effectiveness of the coercive power that rules this society. This is because, as he explains in the *Tractatus theologico-politicus*, a correct appreciation of the extent of the sovereign's rights and power not only includes his ability to compel men by fear, but also takes into account other means by which subjects can be induced to perform certain actions. Actions spring from a deliberation of man with himself, and obedience consists not so much in the outward act as in the mental state of the person obeying: 'Consequently, that ruler has the greatest authority who reigns over the hearts of his subjects'.[92] The preservation of a state depends on the subjects' state of mind, i.e. their 'constancy of heart in carrying out commands'.[93] Since the subjects have transferred many but not all their natural rights to the sovereign, merely coercive power can never be completely effective.[94]

If a well-functioning state depends on the fidelity of its subjects, and if its powers to compel by fear are limited, what other means does it have to curb free-riders? Spinoza appreciates the importance of this problem: 'To prevent all these things, and to establish things so that everyone, whatever his mentality, prefers the public right to private advantage, this is the task, this is our concern'.[95] Given the scope of this problem, it is not surprising that he suggests several solutions.[96] In the *Tractatus theologico-politicus* he suggests a way out that includes an extraordinarily powerful sovereign, but also takes into account the hearts and minds of the subjects. This answer consists of an appeal to transcendent beliefs as exemplified by the Jewish theocracy inaugurated by

91 Rosenthal, 'Action Problems', 392–7; cf. Den Uyl, *Power*, 44–52. For Spinoza the jump from state of nature to civil society is relatively easy because he holds, contrary to Hobbes, that natural law applies in both stages. Natural law applies not only in civil society but already applies in the state of nature. See Spinoza, *Letters*, vol. ii, 406; see also Den Uyl, *Power*, 146–61.

92 Spinoza, *Theological-Political Treatise*, xvii, 297.

93 Spinoza, *Theological-Political Treatise*, xvii, 298.

94 Cf. Armstrong, 'Natural and Unnatural Communities', 293–300.

95 Spinoza, *Theological-Political Treatise*, xvii, 299.

96 See Rosenthal, 'Action Problems', 400–4.

Moses.[97] After their liberation from the bondage of the Egyptians, the Israelites were bound by no covenant to any man. Following the advice of Moses, they decided to transfer their rights to no human being, but only to God. This was a potent answer to free-riding in two ways. Firstly, contrary to a human absolute monarch, God's power to punish (in the present world or in the world to come), and hence his ability to compel by fear, is unlimited. Secondly, their belief in being chosen contributed to 'strengthen the hearts of the Hebrews to bear everything with special constancy and virtue, for the sake of their Country'.[98] The strong collective identity of the Israelites and their belief in collective punishment by God in case of transgressions, curbed individual free-riding. So the Jewish theocracy was very efficient in securing the 'fidelity and constancy in carrying out the orders they receive' of its subjects.

To summarize, for Hobbes, the social contract is the preferred answer to the assurance games in the state of nature and it is certainly preferable to the prisoner's dilemmas in this state. And the power of an absolute sovereign is the only solution to the free-riding that accompanies all civil societies. For Spinoza, on the other hand, the social contract is the exclusive answer to prisoner's dilemmas in the state of nature, but the use of coercive power to curb free-riding in civil society is complemented with suggestions that take into account the psychological state of the subjects. In spite of these differences it is clear that game-theoretical problems are very prominent in the works of both Hobbes and Spinoza. Both regard a transition from the state of nature to civil society as the main or even the exclusive answer to the prisoner's dilemmas that mark this stage, although this creates the challenge of free-rider problems in civil society for which they suggest solutions as well.

Given the importance of game-theoretical problems in De la Court's work, the importance of game-theoretical problems for contract theory, De la Court's acquaintance with the works of Hobbes, and assuming a 'bridge position' between Hobbes and Spinoza, it is remarkable that in the *Maxims* contract theory receives scant attention. One reason for this omission may be De la Court's self-declared aversion to academic and utopian political theories and his preference for practical maxims. He ridicules speculative attempts 'to build *rempublicam Platonis, Aristotelis, eutopiam mori*, a philosophical republick in the air'.[99] In addition, he probably makes a derisive allusion to the *tabula-rasa* moment that is supposed to characterize the contractual inauguration of civil society in contract theory: '[S]o many people cannot be suddenly brought to

97 Spinoza, *Theological-Political Treatise*, xvii, 300–22.
98 Spinoza, *Theological-Political Treatise*, xvii, 314.
99 De la Court, *Maxims*, i.2, 13.

an uninhabited country, to erect a political state … and keep it on foot when it is establish'd. And since in all populous countries there is some form of government; therefore I say again, that those speculations are for the most part useless'.[100] He stresses that politics is not a ball game, that there is never a moment when we 'set the ball fair'—this moment is indeed assumed by political contract theory. Rather, politics demands pragmatism, which means that we should strike the ball 'as it lies'.[101]

De la Court's pragmatic approach did not preclude the presence of an implicit political philosophy in the *Maxims*. This philosophy had been formulated more explicitly in the *Consideratien en exempelen van staat* (fourth edition 1662, 'Considerations and examples of the state') that Pieter had written with his brother Johan and also in the *Politieke discoursen* (second edition 1662, 'Political discourses'). The political philosophy of the De la Court brothers carried Hobbesian overtones, but there were also important differences. For Hobbes, conflicts in the state of nature were primarily conflicts of rights; these conflicts had been resolved after individuals had concluded a covenant and transferred their rights to the sovereign. This model was supposed to work regardless of the precise political form of the power of the sovereign (absolutist or democratic). For the De la Courts, the essential conflict was not between rights but between the passions of self-interested individuals.[102] Since these passions remained as vigorous in civil society as they had been in the state of nature, a transition marked by a covenant was of limited relevance. Statecraft largely revolved around the issue of how to keep the passions of subjects in check, and some forms of government are better at this than others. So whereas the emphasis in Hobbes is on the social contract, in the De la Courts it is on the ideal form of government (democratic).[103]

In addition to this general explanation there is probably a more specific reason for Pieter de la Court's lack of interest in social contract theory in the *Maxims*—which is all the more striking given his interest in contract theory in previous works.[104] I have defended the fruitfulness of a reading of the *Maxims* in strategic terms. It was De la Court's explicit and self-assigned mission to 'enquire into, and lay down some maxims for *Holland's* continual prosperity', both in relation to the other United Provinces and to the other European nations. Since his aims were strategic it makes sense to discuss the *Maxims* in

100 De la Court, *Maxims*, i.2, 13.
101 De la Court, *Maxims*, i.2, 14.
102 Weststeijn, 'Passion', 80–2.
103 See Malcolm, 'Malcolm and Spinoza', 547–9; Kossmann, 'The Course', 74–83.
104 Weststeijn, *Commercial Republicanism*, 142–7.

the strategic terms of game theory. Given this background it is plausible that De la Court did not discuss or develop a social contract theory in the *Maxims*, simply because he felt that this would not contribute to his grand strategy.

We have seen that De la Court's discussion of Holland's relations with the other provinces is expressed in terms of the free-riding behaviour of the latter, who did their utmost to evade the obligations stipulated by the Union of Utrecht. The obvious answer, as formulated by Hobbes, would have been a national sovereign who could have forced the parties to deliver. For De la Court, however, this was neither a realistic nor a desirable option. It was not realistic because under the conditions of the Union of Utrecht political sovereignty remained with the participating provinces and towns. And it was not desirable because the obvious candidate for national sovereign would have been the Prince of Orange, which would have been completely at odds with De la Court's political preferences. So a social contract theory along Hobbesian lines must have shown little promise to De la Court when he tried to grapple with the problem of Holland and its free-riding Dutch allies.

The case of Holland's relations with its European neighbours is even starker. Here a supranational contract not only had little to offer but, from De la Court's point of view, was even directly opposed to Holland's interests. The European nations were in a state of nature that was characterized by constant and chronic mutual warfare. The whole point of De la Court's cat-like strategy was that Holland should not participate in these wars or try to end them, but rather try to profit from these conflicts. Seen from this perspective, an attempt to suggest a 'solution' to the international anarchy would have been self-defeating.

8 Conclusion

The *Maxims* offer a grand strategy for Holland in which a central role was played by free-riding behaviour. The bellicose and intransigent behaviour of the European powers during the General Crisis of the mid-seventeenth century was to be exploited by derived and reversed free-riding, while the free-riding behaviour of the other six allied Dutch provinces was to be curbed. The resulting strategy was coherent because political, military, and diplomatic means should all be used as means towards the same end of commercial prosperity, and also because the relations with both the Dutch allies and the European powers should be dominated by the threat or actual pursuit of the same cat-like opt-out behaviour.

Game-theoretical problems were not only important for De la Court but also played a substantial role in the political philosophy of Hobbes and Spinoza.

Both formulated a social-contract theory as an answer to the prisoner's dilemmas that bedevil the state of nature, and both offer solutions to free-rider problems in the civil society inaugurated by the contract. Since De la Court has been regarded as a link between Hobbes and Spinoza, and since he was just as preoccupied with free-rider problems as these authors, his complete neglect of social contract theory in the *Maxims* is conspicuous. This omission becomes understandable once it is appreciated that the *Maxims* do not aim to present an abstract political philosophy but a practical grand strategy that had little use for contract theory.

Until the military disasters of 1672, De la Court's *Maxims* were in many ways not only a prescription but also the description of a remarkably successful strategy. Yet in the later wars against Louis XIV, when the Dutch Republic and England became partners—with the Dutch fighting under general Marlborough while paying for most of the war effort against the French, and with the English taking over maritime supremacy—the cat's grand strategy was unravelled by a set of policies and circumstances that confirmed De la Court's worst nightmares.

CHAPTER 2

François Fénelon on Luxury, War, and Trade in the *Telemachus*

1 Introduction

In his novel *Les aventures de Télémaque* (*Telemachus, son of Ulysses*), published in 1699, François de Salignac de la Mothe-Fénelon (1651–1715) expresses strong and original views on the topics of luxury, war, and trade.[1] His negative views on luxury, his critical evaluation of war and trade, and his analysis of the relation between these phenomena would resonate through the entire eighteenth century. He would influence Shaftesbury, Hutcheson, Berkeley, and Rousseau. His critics, among whom were Mandeville, Montesquieu, and Voltaire, would put forward a more positive evaluation of luxury and trade, and would connect these phenomena with peace rather than war.[2] I will discuss the vital and often paradoxical relation between these three elements in Fénelon and sketch the background to his views, with special attention to the military expansion and the mercantilism of Louis XIV, Fénelon's quietist spirituality, and the development of the concept of self-interest in seventeenth-century philosophy by mechanical philosophers and economic thinkers.[3]

2 Telemachus

Fénelon was tutor to the Duke of Burgundy, the eldest son of the *Dauphin,* between 1689 and 1697. His rise in the favour of King Louis XIV culminated with his consecration as archbishop of Cambrai in 1696, but was followed by a sudden disgrace only one year later, largely due to his role in the Quietist

1 Fénelon, *Les Aventures de Télémaque*, ed. Albert Cahen, 2 vols (Paris: Hachette, 1920); translation: Francois de Fénelon, *Telemachus, son of Ulysses*, ed. and transl. by Patrick Riley (Cambridge: Cambridge University Press, 1994).

2 See Berry, *The Idea of Luxury*; Hont, 'The Early Enlightenment Debate', 379–418; Jennings, 'The Debate', 79–105; Wahnbeck, *Luxury.*

3 For Fénelon and luxury see Bonolas, 'Fénelon et le luxe', 81–90; on war see Lorson, 'Guerre', 207–14; on commerce see Hont, *Jealousy of Trade*, 9 and 25. I have found no study devoted to Fénelon and the relation between these topics.

 | DOI:10.1163/9789004536678_004

controversy around Jeanne Marie Bouvier de la Motte Guyon (1648–1717). In 1694–1696 he wrote a novel for his royal pupil, the *Telemachus*. Its publication in 1699, although possibly without Fénelon's explicit permission, confirmed Louis XIV's decision to prolong Fénelon's banishment from the court of Versailles.[4] The novel is set in a fictional Homeric world and contains several longer passages that have a clearly utopian character.[5] At the same time the *Telemachus* is an example of the *specula principium* or 'mirrors for princes' that had been a popular genre since the Renaissance.[6] While Telemachus is traversing a pseudo-Classical Mediterranean world in search of his father Ulysses after the fall of Troy, the young prince is constantly lectured about his future duties as sovereign of Ithaca by the goddess Minerva, who has taken the human form of the wise Mentor, and whose ulterior *alter ego* is not Minerva but Fénelon himself.[7]

Whilst recounting Telemachus' peregrinations, Fénelon takes ample time to denounce the evils of luxury and war. Even the best characters are corrupted by luxury, softness (*molesse*) and idleness.[8] If people are allowed to live in abundance, they will stop working, become proud, undisciplined, and apt to revolt.[9] A good king curbs excessive wealth, reduces pomp and forbids shameful pleasures. He will only support arts that help to meet true needs and he will prohibit those that inspire vice.[10] When Telemachus courageously descends into the depths of Hades in a vain search for his father, he meets several former kings of Lydia, who are all being punished because they preferred luxury to a life of work in the service of their people.[11]

Fénelon's condemnation of war is as scathing as his attack on luxury. Even if some wars cannot be prevented, and even if some are necessary to defend a nation, they are always a shame to the human race. Men are all brothers and they should love each other rather than tear each other to pieces.[12] States

---

4 Cahen, introduction in Fénelon, *Télémaque*, vol. 1, li–liii.

5 Utopian texts around 1700 had their roots in various *genres*, such as *romans archéologiques*, *arcadies*, *uchronies*, *robinsonades* and imaginary voyages. See Cahen, introduction in Fénelon, *Télémaque*, vol. 1, xxviii–xxxiv, lxviii; Racault, *L'Utopie*, 182; Trousson, *Voyages*, 97.

6 Bonolas, 'Fénelon', 81; Racault, *Utopie*, 192.

7 Cf. Shklar, *Men and Citizens*, 4, who argues that the genres of the utopia and the 'mirror of princes' are actually quite compatible because they allow us 'to judge the actual by confronting it with the perfect'.

8 Fénelon, *Télémaque*, vol. 2, xi, 188.

9 Fénelon, *Télémaque*, vol. 2, xi, 161.

10 Fénelon, *Télémaque*, vol. 1, v, 189; Fénelon, *Télémaque*, vol. 1, v, 220–1.

11 Fénelon, *Télémaque*, vol. 2, xiv, 340.

12 Fénelon, *Télémaque*, vol. 2, viii, 48 and Fénelon, *Télémaque*, vol. 2, xiii, 281.

are exhausted by wars, which put their very survival in jeopardy.[13] Triumphant conquerors have the dubious majesty of large rivers that have broken through their dikes, ravaging rather than irrigating the countryside. Conquerors are sent by the gods to punish men, and acquit themselves from this task by destroying kingdoms, sowing fear, misery, and desperation, and enslaving free men.[14] Even victorious wars tend to have nefarious consequences, if only because the victors will tend to quarrel over the spoils.[15]

The background to Fénelon's criticism of luxury and war is rich and complicated, but the most obvious context is formed by the internal and external politics of King Louis XIV. The *Telemachus* is not a direct satire or a *roman à clef*, but Fénelon's verdict on wars waged by ambitious kings, who are driven by egoistic pride and who devastate countries for '[a] false glory, a vain title of conqueror' implies a clear condemnation of Louis XIV's expansionist wars.[16] When Mentor explains to Telemachus that before a war is started the first question should be about its justice, he clearly implies the sad absence of this commodity in the considerations of the Sun King.[17] Similarly, his condemnation of luxury is directed against an economic and fiscal policy that favours the production of luxury goods supporting the decadent court of Versailles and the rich inhabitants of Paris and other large cities, at the expense of the backbreaking work of the denizens of the countryside. Hence Mentor can exclaim 'that there are two grievances in government which are scarcely ever guarded against or remedied: the first is an unjust and violent authority assumed by kings; the second is luxury, which corrupts manners'.[18]

It has been noted that Fénelon's complaints about social inequality, luxury, the misery of the common people, the disasters of war and its false glory fit a broad tradition of both secular and Christian moralists.[19] Fénelon's views in the *Telemachus* had an added poignancy because they were proposed to a pupil that had been entrusted to him by the King himself. Moreover, Fénelon's critical allusions to the policies of his monarch in the *Telemachus* are matched and confirmed by views expressed in other writings that are even more explicit—both on war and on luxury.

13 Fénelon, *Télémaque*, vol. 2, xi, 189.
14 Fénelon, *Télémaque*, vol. 1, vii, 334.
15 Fénelon, *Télémaque*, vol. 2, xvii, 470; see also Fénelon, *Examen de conscience*, ii, 991.
16 Cahen, introduction in Fénelon, *Télémaque*, vol. 1, xxxix; Fénelon, *Telemachus*, 230/ Fénelon, *Télémaque*, vol. 2, xiii, 282.
17 Fénelon, *Télémaque* vol. 2, viii, 4–5.
18 Fénelon, *Telemachus*, 296/*Télémaque* vol. 2, xvii, 464–5.
19 Cahen, introduction in Fénelon, *Télémaque*, vol. 1, xxv–xxvi.

In his *Mémoire sur la campagne de 1712* ('Note on the campaign of 1712') Fénelon wrote, 'It would not be fitting for me to reason on war and I have taken care not to commit such an absurdity'.[20] But in 1701 he had already written a *Mémoire sur les moyens de prévenir la guerre de la succession d'Espagne* ('Note on the ways to prevent the Spanish war of succession'), which is actually an essay on French military grand strategy. Between 1701 and 1712 Fénelon wrote nine papers on the War of the Spanish Succession and its aftermath. These papers offer a good view of the war from a French perspective, although this certainly was not a French perspective that King Louis XIV himself viewed with much sympathy. Fénelon's paper from 1701 can be read as a pragmatic argument in favour of a limited war.[21] Subsequent papers, written between 1710 and 1712, had a more outspoken character. By 1710 the war seemed lost and there was concern that its continuation would put the very survival of the French nation at stake. Fénelon's arguments for a limited war changed into a passionate plea for peace at almost any price. The catastrophic state of the war-weary French economy since 1708, the difficulties of maintaining the army, the internal disorder and misery were all used to stress this ardent wish.[22] In this context Fénelon repeatedly stressed the importance of renunciation. The only glory that the French could wish for the French King was that he should turn his courage against himself and sacrifice his military ambitions to save the country that he had received from God. In the same spirit his grandson Philip V should renounce his disputed Spanish throne. For Fénelon, this renunciation was dictated by the law of nations, by the European balance of power, and by peace itself.[23]

Louis XIV also provided a context to the topic of luxury. In his essay on the duties of a king, Fénelon asks 'Do you take care to repress luxury?'[24] And in his advice on the education of the daughter of 'a lady of quality' he stresses that it is 'luxury that confounds all situations, lifts up people of low birth, who are hastily enriched by detestable means above people of the most elevated position. Moreover, luxury causes a chaos which corrupts the morals of a nation, arouses greed, makes one used to intrigues and acts of meanness, and gradually

20 Fénelon, *Mémoire sur la campagne de 1712*, ii, 1075.

21 Fénelon, *Mémoire sur les moyens de prévenir la guerre de la succession d'Espagne*, ii, 1013–27.

22 Fénelon, *Mémoire sur la situation déplorable de la France en 1710*, ii, 1034.

23 Fénelon, *Examen des droits*, ii, 1069; see also the famous but unpublished *Lettre à Louis XIV*, I, 541–54; Racault, *Utopie*, 202. See below, §3.4 for Montesquieu on the plans of Louis XIV for a universal monarchy.

24 Fénelon, *Examen de conscience*, ii, 980.

undermines all the foundations of decency'.[25] Finally, in the so-called *Tables de Chaulnes* of 1711 Fénelon sketches a reform programme in which peace with Europe, reductions of the budgets for army, navy, and the court, and sumptuary laws reminiscent of the Roman Republic, figure prominently.[26]

In the *Telemachus* Fénelon sketches what Istvan Hont calls a 'tripartite model of the history of luxury': Boetica, unreformed Salentum and reformed Salentum.[27] Many aspects of Fénelon's social, economic, and political philosophy are represented in this model as well, including his analysis of war and trade.

3 Boetica

In book VII of the *Telemachus* a Phoenician ship helps the eponymous hero to escape the lures of Calypso. Adoam, the brother of the captain, informs Telemachus and Mentor about earlier voyages. Adoam's most curious tale is about the primitive and isolated tribe of the Boeticans, who are reported to live in Spain, at the river that is presently called the Guadalquivir.[28] They are engaged in agricultural and pastoral occupations. Thanks to their frugality, moderation, and pure morals they enjoy a long life free of diseases. They have no cities and do not even need houses: 'Every family moves from one part of this charming country to another, after having consumed the fruits and pasturage of the place where they had pitched their tents'.[29] They dislike all things that 'enervate, intoxicate, and torment those who possess them'.[30] They use no iron except for the most basic agricultural tools, and they detest all the arts that are held in such high regard by the Greeks and the Egyptians. Since each family is a self-sufficient economic unit, there is hardly any need for mutual barter and certainly no need for gold or money. The Boeticans have all their property in common. Although they do not need imported goods, they are so hospitable as to allow the Phoenicians a trading post, and they give these merchants freely what they do not need themselves, without any payment in return: 'They take pleasure in giving away their superfluities to strangers'.[31] They are extremely

25 Fénelon, *Avis de M. de Fénelon*, ii, 1132.
26 Fénelon, *Tables de Chaulnes*, ii, 1085–1105; cf. Fénelon, *Télémaque*, vol. 2, x, 191 n. 1.
27 Hont, 'The Early Enlightenment Debate', 384.
28 Fénelon, *Télémaque*, vol. 1, vii, 320–42.
29 Fénelon, *Telemachus*, 111/Fénelon, *Télémaque*, vol. 1, vii, 331.
30 Fénelon, *Telemachus*, 110/Fénelon, *Télémaque*, vol. 1, vii, 328.
31 Fénelon, *Telemachus*, 113/Fénelon, *Télémaque*, vol. 1, vii, 340.

peaceful, both amongst each other and with neighbouring tribes, which are allowed to take the lands of the Boeticans, as long as enough land remains for their own purposes. While they refuse to dominate others, it is a well-known fact that they would rather die than be enslaved by other people. All this permits the Boeticans to live in profound peace with all their neighbours.

Boetica allows Fénelon to expose an exemplary absence of luxury, war, and commerce. The frugality and the shared property of the Boeticans guarantees an absence of internal strife, while their modest standard of living robs their neighbours of a motive to acts of aggression. In addition, a lack of luxury goods implies minimal trade; and since the Phoenician traders are given freely what they desire, they have no reason to start a commercial war against the Boeticans. If a pastoral life-style ensures frugality, internal and external peace, the slaughter of live-stock seems an almost inevitable 'collateral' consequence of this life-style. Even here, however, Fénelon takes care to maintain a pacifist stance: 'Never did human blood stain the land, and even that of lambs but seldom'.[32] The Boeticans prefer the fruit of the trees and the milk of their herds and flocks. So, in Boetica the absence of luxury, violence, and commerce form the interconnected elements that stabilize and maintain a society that is the precise opposite of the France of Louis XIV. Adoam's account of the Boeticans is utopian and wildly unrealistic, but it has a clear function in the *Telemachus,* whose hero describes it as a 'belle fable'.[33] Boetica does not function as the first phase of a descriptive socio-economic history, but rather as a normative paradigm.

4 Salentum: Luxury and War

In book VIII Telemachus and Mentor visit Salentum (Italy), ruled by King Idomeneus. Salentum is the very antithesis of idyllic Boetica. Idomeneus has started a war against the Mandurians, and he has indulged in the luxury of great but useless building projects. The relation between the *presence* of both luxury and war in Salentum is as clear as the relation between the *absence* of both luxury and war in Boetica. In Salentum, pomp, splendour and other excesses have thrown men into a violent state that tempts them to break the law in order to satisfy their inordinate desires.[34] The relation between luxury and Salentum's trade is more ambiguous; while on the one hand luxury should

32 Fénelon, *Telemachus*, 111/Fénelon, *Télémaque*, vol. 1, vii, 333.
33 See also Racauld, *Utopie*, 197–8.
34 See also Fénelon, *Télémaque*, vol. 2, xiv, 342.

have acted as a stimulus, the increasing fiscal burdens used to pay for the king's wars have actually stifled trade.[35]

In book x Idomeneus delegates to Mentor the conception and execution of a far-reaching reform programme, which is brought to completion in book XVII. While Boetica was meant as a primitive utopian ideal, Mentor's work in Salentum is supposed to provide a practical reconstruction programme for contemporary France, which was commercially well-developed and fully monetized and which suffered under the burden of Louis XIV's hegemonic military policy.[36] Once Mentor has extricated Salentum from its war with the Mandurians, he vigorously proceeds with a massive assault on luxury and all its perceived concomitant evils. His aim is to restore the kingdom to 'a noble simplicity and frugality'.[37] A king's strength is determined by the number of good citizens and the size of the well-cultivated lands needed to feed them.[38] The king's interests are not served by a superb city in marble and gold in the midst of an exhausted countryside, but by a flourishing agricultural sector and a small city with virtuous inhabitants. In order to achieve this aim, Mentor is prepared to take radical measures.

All superfluous city-dwellers of Salentum are 'transplanted' to the countryside.[39] The city becomes an unadorned 'solitude' where the arts languish.[40] Each citizen receives his own plot of land, which however is so small that he is well motivated to work it extensively.[41] Besides being the only true source of wealth, agriculture is also a source of social stability; constant toil deflects farmers from attempts to subvert the social order.[42] Once people are allowed to cultivate the land in peace the spectre of luxury might again raise its head, but all 'superfluous arts, which divert the poor from the culture of the lands for the supply of real needs, and corrupt the rich by introducing among them luxury and softness' have been banished.[43] Reformed Salentum society is heavily regulated and rigidly stratified. Each of the seven fixed social classes has its

35 Fénelon, *Télémaque*, vol. 2, x, 61–7; see also Fénelon, *Télémaque*, vol. 2, x, 66 n. 3 on the decline of trade under the burdens of war under Louis XIV.
36 Cf. Jasinski, 'Sur Fénelon', ii, 271.
37 Fénelon, *Telemachus*, 163/Fénelon, *Télémaque*, vol. 2, x, 99; see also Fénelon, *Télémaque*, vol. 1, v, 220–1.
38 Fénelon, *Télémaque*, vol. 2, x, 62.
39 Fénelon, *Télémaque*, vol. 2, xvii, 463; cf. Fénelon, *Télémaque*, vol. 2, x, 112.
40 Fénelon, *Télémaque*, vol. 2, xvii, 461.
41 Fénelon, *Télémaque*, vol. 2, x, 121.
42 Cf. Racault, *Utopie*, 201–2.
43 Fénelon, *Telemachus*, 296/Fénelon, *Télémaque*, vol. 2, xvii, 464.

own distinctive dress.[44] Public schools install patriotism and respect for the laws, and the pupils are taught to put honour above pleasure and life itself.

Since Fénelon supposes that luxury encourages soft and effeminate behaviour, it is not surprising that he posits a negative relation between luxury and a people's capability to wage war. Indeed, Fénelon clearly associates luxury with cowardliness, in the *Telemachus* and elsewhere.[45] In one of Fénelon's *Dialogues des morts*, King Pyrrhus of Epirus blames King Demetrius I Poliorcetes of Macedonia for his softness and cowardliness during peace, but Demetrius responds that *in spite* of 'a little softness' he has been a great soldier, which supposes again an (implicit) negative relation between luxury and military prowess.[46] The perception of a similar relationship in the eighteenth century would influence subsequent debates about the nature and causes of war. Montesquieu and others believed that the increased wealth and mutual commercial interdependence of the European powers made them less prone to military adventures. In the period between the death of Louis XIV and the wars of the French Revolution and Napoleon, this hypothesis seemed to be confirmed by a real trend.[47]

These considerations lead to an intriguing paradox in Fénelon's views. On the one hand he rejects both luxury and war, but on the other hand there is the possibility of an inverse causal relationship between the two phenomena—which might have brought him to a more positive evaluation of luxury. This paradox brings him to a more detailed analysis that leaves his negative appraisal of luxury completely unchanged, but prompts him to an evaluation of war that is considerably richer than might be expected. He starts with the observation that if luxury negatively affects the ability to wage war, this is merely the effect of effeminate cowardliness, which is not something that should be encouraged.[48] A good king desires peace because of his wisdom and moderation, not because of his preference for a soft life or fear of war.[49] To fear war out of softness is as reprehensible as waging war out of ambition.[50] Yet although Fénelon believes that luxury decreases the *capability* to wage war, contrary to later thinkers he does not think that it contributes towards peace. On the contrary: the envy and restlessness associated with luxury seem to increase the chance of an *actual*

44 Racault, *Utopie*, 201.
45 Fénelon, *Télémaque*, vol. 1, iv, 164–5.
46 Fénelon, *Dialogues des morts*, i, 368.
47 Rahe, *Soft Despotism*, 27–42; see also below §3.4.
48 Fénelon, *Télémaque*, vol. 1, iv, 164–5.
49 Fénelon, *Télémaque*, vol. 2, viii, 37.
50 Fénelon, *Télémaque*, vol. 2, xi, 192.

outbreak of armed conflict. Indeed, for a Frenchman in the final decades of Louis XIV's reign, appreciation of a trend towards a decrease in the frequency and seriousness of military aggression, caused by whatever factor, must have been extremely difficult. Moreover, Fénelon's account of Salentum, however wildly improbable it may read, was meant as a realistic programme that would prepare his royal pupil for a future that would hopefully remain untainted by aggressive aspirations to a universal monarchy. The new king would be fully entitled to wage defensive wars. In these wars luxury was perceived not as an asset but rather as a liability.[51]

So Fénelon's views on warfare are complex, and in spite of pacifist sympathies, the *Telemachus* is partly, in the words of Albert Cahen, 'a small treatise on military courage and the duties of a commander-in-chief'.[52] While in the previous chapter we saw De la Court advocate the cautious strategy of the cat, who is even prepared to accept blackmail if this contributes to his chances of survival, Fénelon definitely admires the pugnacious temper of the royal lion. A king who only knows how to reign during peace is only half a king.[53] If war cannot be avoided, then the place of a king should be at the head of his armies.[54] For Fénelon, one of the great functions of a king is still to defeat his enemies.[55] When Telemachus assists Idomeneus in a defensive war against King Adrastus, he receives the following encouragement by Mentor: 'Do not, therefore, O Telemachus, avoid any danger, but rather lose your life than have your courage called into question'.[56] Hence care should be taken that luxury will not cause a nation to fall into ignorance about the art of war.[57]

Fénelon not only makes the negative point that luxury affects military capabilities, but he even makes the more positive point that preparing for war stimulates civic virtues and serves as a disincentive to luxury. Children should be taught courage, which does not consist solely 'in despising death amidst the dangers of war, but also in disdaining excessive wealth, and shameful pleasures'.[58] The glory and virtue of heroes should be chanted, trophies should be awarded in order to stimulate emulation, and in public schools youths should

51 See Fénelon, *Télémaque*, vol. 1, vii, 334; Fénelon, *Télémaque*, vol. 2, viii, 48; Fénelon, *Télémaque*, vol. 2, xiii, 283. Compare with Montesquieu's critique of universal monarchy, below, §3.4.

52 Cahen, introduction in Fénelon, *Télémaque*, vol. 1, xxx.

53 Fénelon, *Télémaque*, vol. 1, v, 217.

54 Fénelon, *Examen de conscience*, ii, 976.

55 Fénelon, *Télémaque*, vol. 1, v, 222–3.

56 Fénelon, *Télémaque*, vol. 2, x, 71.

57 Fénelon, *Télémaque*, vol. 2, x, 68.

58 Fénelon, *Telemachus*, 60/Fénelon, *Télémaque*, vol. 1, v, 189.

be accustomed to the most rigorous exercises.[59] This programme of exercise agrees with Fénelon's preference for a society consisting of small farmers who are used to physical hardship and are prepared to defend their country with their own hands. He is inspired no doubt by the example of the early Roman Republic; and in one of Fénelon's *Dialogues des morts*, Cato the Elder boasts 'But do not believe that this application to agriculture and eloquence has diverted me from the art of war'.[60]

So far, Fénelon's views can be understood in terms of a well-known connection between a preparedness for war that includes civic virtue and frugality, and a resulting avoidance of actual military conflict, along the lines of the classic dictum *si vis pacem, para bellum*. He writes indeed: 'The surest way to prevent war and to secure a long-lasting peace is to have your people trained in the art of combat'.[61] He even seems prepared to go one step further. The only way to ensure that a nation's young aristocracy does not languish in a perpetual peace, forgets the art of war, and becomes affected by luxury, is its actual *participation* in wars.[62] At the same time, the flower of Salentum's youth should not be mobilized for wars by the state itself, but sent to assist its allies in their wars. In that way, Salentum will always be able to call upon the services of 'a gallant and intrepid youth', avoid a war at its own cost and peril, preserve its high reputation amongst its allies, and be able to count on their assistance when it is attacked by an aggressor. Thus ultimately the active participation of Salentum's young elite in wars abroad is a further instance of prevention of war by preparing for war.[63]

Mentor's defence of this training programme is put forward within a context formed by the war of Salentum and its allies against the aggressive campaign of Adrastus. So here we see a discussion of a defensive war in which the allies try to contain an expansionist adversary and thus contribute to the balance of power. The importance of an international balance of power is alluded to in the *Telemachus*.[64] Fénelon discusses the balance of power more explicitly in an essay on the duties of a king. In this essay he stresses the importance of maintaining tranquillity through participation in international alliances against nations who attempt to establish a universal monarchy.[65] This

59 Fénelon, *Télémaque*, vol. 2, xi, 188.
60 Fénelon, *Dialogues des morts*, i, 391; see also Fénelon, *Télémaque*, vol. 2, xvii, 464.
61 Fénelon, *Telemachus*, 196/Fénelon, *Télémaque*, vol. 2, xi, 191.
62 Fénelon, *Télémaque*, vol. 2, xI, 189; Fénelon, *Télémaque*, vol. 2, x, 68.
63 Fénelon, *Telemachus*, 197/Fénelon, *Télémaque*, vol. 2, xi, 191.
64 Fénelon, *Télémaque*, vol. 2, viii, 39.
65 See also Montesquieu, *Réflexions*, note by Catherine Larrère and Françoise Weil, 363 n. 66: 'Pour Fénelon comme pour Leibniz, le maintien de l'équilibre justifie la formation

contribution by young nobles to alliances that are used to balance aggressive attempts at universal monarchy is a remarkable idea in a novel written for the Duke of Burgundy, since at the time the *Telemachus* was produced, the only monarch perceived to be aiming for universal domination was the Duke's own grandfather, Louis XIV.

The perceived value of a military education, both as a preparation for justified defensive wars and as a disincentive to luxury, explains Fénelon's praise of male and martial virtues, in spite of his fundamental pacifism. Mentor's voice is said to lack any feminine softness and when he actually participates in combat, his eyes glow with an audacity that surprises the fiercest combatants.[66] Later he is emulated by his pupil Telemachus, who shows courage in the midst of military dangers, and whose face on the day of battle with Adrastus radiates with a proud majesty that promises victory.[67] Moreover, Fénelon shows an appreciation of male beauty which rather surpasses his description of the charms of the young nymph Eucharis, who besotted Telemachus on the treacherous island of Calypso, and which certainly overshadows his rather bland sketch of Telemachus' prospective bride Antiope, who is modestly characterized as 'good-natured, discreet, and unaffected'.[68] Fénelon's admiration of male beauty is expressed in terms that confirm his aversion to effeminate luxury and his positive evaluation of martial strength. Telemachus has 'something very sprightly and amiable in his air and countenance; and his person is adorned with all the graces of youth and beauty; but it is a beauty neither languid nor effeminate: even in the tender blossom of early youth, he appears vigorous, hardy, and robust'.[69] And when Telemachus fights Iphicles, son of Adrastus, we are informed that the combatants are 'beautiful, strong, active, and courageous'.[70]

d'alliances "tans offensives que défensives"'; see also Malettke, 'Fénelon', 480; Hanley, *The Political Philosophy*, 31.

66 Fénelon, *Télémaque*, vol. 1, vii, 316; Fénelon, *Télémaque*, vol. 1, i, 40.

67 Fénelon, *Télémaque*, vol. 2, xii, 197; Fénelon, *Télémaque*, vol. 2, xv, 398.

68 Fénelon, *Telemachus*, 303/Fénelon, *Télémaque*, vol. 2, xvii, 483; see also Hillenaar, *Le secret*, 45: 'Nous apprenons seulement "qu'elle était vêtue comme Diane. Vénus et Cupidon avaient répandu sur elle de nouveaux charmes", mais ces charmes, c'est au lecteur de les deviner. L'auteur s'arrête en revanche longuement et avec beaucoup de détails physiques aux tourments du jeune homme amoureux. Sa beauté est décrite avec une métaphore qui est traditionnellement réservée à l'évocation d'une femme ou d'un enfant'.

69 Fénelon, *Telemachus*, 123/Fénelon, *Télémaque*, vol. 1, viii, 366; see also Hillenaar, *Secret*, 50: 'la peur, qu'on trouve un peu partout dans Télémaque, d'être "efféminé"'.

70 Fénelon, *Telemachus*, 228/Fénelon, *Télémaque*, vol. 2, xiii, 278; see also Fénelon, *Télémaque*, vol. 2, xiii, 300; Fénelon, *Télémaque*, vol. 2, xv, 414–5. On Fénelon and 'true courage' see Hanley, *The Political Philosophy*, 83–115.

5 Salentum: Commerce

So far, we have concentrated on the relation between war and luxury in reformed Salentum; let me now turn to the third coefficient of the equation: commerce. Fénelon's point of departure seems to be freedom of trade. When Mentor embarks on his reform-programme of Salentum, he advises Idomeneus to invite back the foreign merchants who had been chased away by excessive customs duties.[71] Foreigners should be allowed to settle in Salentum and merchants who bring the commerce of foreign nations to Salentum are even provided with a bonus.[72] Fénelon probably implied that France should open its ports to foreign, especially Dutch, merchants. His description earlier in the *Telemachus* of Tyre, probably modelled on contemporary Amsterdam, can be read as praise for the Dutch spirit of enterprise.[73] In practise, however, Mentor heavily regulates the trade of Salentum in a variety of ways and for various reasons. These regulations throw light on Fénelon's views on the relation of trade with luxury and war. Unsurprisingly, all foreign merchandise that could introduce luxury and softness is strictly forbidden.[74] The medium of trade is money, and this is a merely artificial form of wealth that diverts people from real work and enables them to indulge in dubious pleasures.[75] The proceeds of the limited trade of Salentum with the outside world should only be used to purchase necessary goods that cannot be produced or obtained within the country itself. Money can also be used to pay for inevitable wars that must be waged abroad.[76] The limited role assigned to money in Salentum's military policy is probably linked to a preference for a civic militia of farmers rather than a mercenary army, and is in line with Fénelon's admiration for the early Roman Republic. There had been many precedents in early-modern military writing for this appreciation, at least since Machiavelli's *Arte della guerra* (1521). These feelings would gain renewed prominence among eighteenth-century *philosophes*, who would consider a citizen army a safeguard against royal tyranny.[77]

71 Fénelon, *Télémaque*, vol. 2, x, 66; Fénelon, *Télémaque*, vol. 1, iii, 120.

72 Fénelon, *Télémaque*, vol. 2, x, 88; Fénelon, *Télémaque*, vol. 2, xvii, 463.

73 Fénelon, *Télémaque*, vol. 1, iii, 111–28; see also Fénelon, *Mémoire sur les moyens de prévenier la guerre de la succession d'Espagne*, ii, 1014. On Tyre/Amsterdam see Hanley, *The Political Philosophy*, 61–6.

74 Fénelon, *Télémaque*, vol. 2, x, 90; Fénelon, *Télémaque*, vol. 2, xiv, 367–8.

75 Fénelon, *Télémaque*, vol. 2, xiv, 370; Fénelon, *Télémaque*, vol. 2, xvii, 462; see also Fénelon, *Examen de conscience*, ii, 1009.

76 Fénelon, *Télémaque*, vol. 2, xiv, 366–9; see also Sonenscher, *Before the Deluge*, 209.

77 See also Palmer, 'Frederick the Great', 107–8.

The connection between limitations on trade and limited economic and military aims is fairly obvious. More remarkable is the positive role that Fénelon sees for trade in curbing rather than stimulating luxury. Mentor advises Idomeneus to exchange all superfluous goods 'that were no longer tolerated' in Salentum, with the neighbouring Peucetians for livestock.[78] This suggests that trade should be used by Salentum as a means to remove luxury goods from society, i.e. trade and luxury seem to be placed in an *inverse* relationship. This is an intriguing position that is at odds with the view taken by subsequent participants in the eighteenth-century luxury debate, in which trade and luxury would often be regarded as two mutually reinforcing phenomena. Fénelon may have hinted at a similar anti-luxury function of trade earlier in the *Telemachus*, in his discussion of the Boeticans for whom, as we have seen, it was a pleasure to give Phoenician traders their superfluities. In more developed and monetized Salentum, giving away is not a realistic option; moreover, we have seen that a limited amount of money should be earned to pay for basic imports and the military campaigns that are forced on Salentum. If an important function of trade is to combat luxury by siphoning off wealth, then it is not necessary to use the nation's own traders. In so far as native merchants would use their profit to import or stimulate the manufacture of luxuries, this would actually be counter-productive to Fénelon's specific aims. So his passive toleration of commerce, as opposed to its active stimulation, and his simultaneous appreciation of Dutch commercial activity, may actually have been inspired by its perceived role in his anti-luxury programme. This could be the background for his remarkably apodictic recommendation in the *Tables de Chaulnes* to 'leave the profit of their austere frugality and their labour to the Dutch', which has the added implication that the Dutch had a culture that made them less prone than the French, or at least the French court, to the conversion of profit into luxury.[79] Fénelon's views in the same *Tables* on the French fleet, which should be 'mediocre, without being pushed to excess', also accords with the idea of a merely passive toleration of trade.[80]

6 Mercantilism

The obvious context for Fénelon's ambiguous and circumspect views on trade is formed by the mercantilist policies of Louis XIV. Fénelon belonged

78 Fénelon, *Télémaque*, vol. 2, x, 131.

79 Fénelon, *Tables de Chaulnes*, ii, 1104.

80 Fénelon, *Tables de Chaulnes*', ii, 1105.

to the circle of the Duke of Beauvillier and the Duke of Chevreuse, who were involved in a futile conspiracy against the influence of the Le Tellier family and their mercantilist policy.[81] Fénelon's negative appreciation of mercantilism informed his implicit assumptions and his explicit criticisms on the relationship between trade and war. Early modern Europe experienced a military development in which firearms transformed both field and siege warfare, and in which armies and navies grew in size and became more permanent. Money used for increased military spending increased the power of the central state, giving it a monopoly of legitimate force and making warfare all the more a state affair, which in turn further increased the state's power of taxation and command.[82] At the same time the increased monetary needs of the early modern state constantly threatened to overwhelm its still cumbersome fiscal apparatus and its fledgling credit facilities, except in the Dutch Republic and England. The accumulation of bullion through a positive trade balance was regarded as an instrument to boost state income, and this entailed a mercantilist programme that stimulated domestic manufactures and export trade and penalized imports with high customs duties.[83] The needs of warfare prompted even the most traditional territorial monarchies to occupy themselves with commerce and resulted in a growing tendency to view trade in a context of opposed interests, or 'jealousy of trade', in policies that stressed national autarchy, i.e. independence from imports, which often left increased xenophobia in its wake.[84]

All these elements were perceived and attacked by Fénelon. His criticism of money as a merely artificial source of wealth, and his preference for agriculture over trade, form an obvious counterpoint to the monetarist and commercial preoccupations of Colbert's mercantilism. Jealousy of trade was in many ways a continuation of Machiavellian reason-of-state policies with economic means, and Fénelon viewed Colbert's policies as the economic side of Louis XIV's 'Italian policy'.[85] Although his evaluation of the mercantilist model was negative, he shared the assumptions about the relation between its main parts. Since mercantilist thinkers and decision makers viewed trade and the resulting financial gains as a means to fuel a nation's military apparatus, it is not

81 Rothkrug, *Opposition to Louis XIV*, 258.
82 Gat, *War*, 443–511; see also above, §1.3.
83 North, *The Rise*, 91–156; Mokyr, *The Lever of Riches*, 78–9; Christian, *Maps of time*, 316–24, 393–401; Gat, *War*, 506.
84 Silberner, *La guerre*, xxvii–lviii; Hont, *Jealousy of Trade*.
85 Hont, *Jealousy of Trade*, 9, 25.

surprising that the role assigned by Fénelon to trade and money is narrowly circumscribed.

This background also explains Fénelon's ambiguous views about the so-called liberty of trade. Insofar as his proposals for freedom of trade implied opening French ports to Dutch traders he directly rejected the mercantilist apparatus of customs barriers that were erected against imports from abroad. This plea for liberty of trade was the expression of opposition against a *specific* form of economic state interference that favoured the accumulation of bullion at the expense of all other interests, including the agricultural interests of the aristocracy.[86] This specific point implied no rejection of state interference *in general*. Fénelon subscribed to the mercantilist assumption that commercial prosperity was foremost the result of international and maritime trade in luxury goods, and he also agreed that this trade should be subject to tight state regulation—not, however, to stimulate but rather to curb this form of commerce.

Fénelon also addressed the autarchic tendencies inherent in mercantilism. He favoured an agricultural society with largely self-sufficient farmers. Yet this agricultural autarchism is different from the autarchic aims of mercantilist policies, which tried to reduce home consumption and increase national economic production in order to achieve a trade surplus that enlarged the state budget. Whereas mercantilist autarchy figured as a strategic weapon in Louis XIV's expansionist foreign policy, Fénelon's autarchy is the attribute of a peaceful socio-economic order that needs little of the goods that are owned by its neighbours and which itself has little that can be exported or might provoke their greed.[87]

Finally, in his essay on the duties of a king, Fénelon observes that countries bound together by trade form 'a great body and a community of sorts'.[88] Yet he shares the antagonistic and mercantilist assumptions about the relatively fixed amount of total wealth that can be derived from trade, which means that one nation's gain is automatically another nation's loss.[89] In the *Telemachus*, in a description of Tyre/Amsterdam, it is remarked that 'the Phoenicians have made themselves the masters of the whole commerce of the world, thus enriching themselves at the expense of all other nations'.[90] Given these antithetic assumptions of mercantilist autarchy, it is not surprising that

86 Rothkrug, *Oppostion*, 433; Mousnier, 'Les idées politiques', 195.

87 Silberner, *La guerre*, xxxiv–xxxvii.

88 Fénelon, *Examen de conscience*, ii, 1005.

89 Silberner, *La guerre*, xxxiv–xxxvii.

90 Fénelon, *Telemachus*, 36/Fénelon, *Télémaque*, vol. 1, 115; cf. Bonolas, 'Fénelon', 84.

mercantilist policies were often marked by xenophobic misgivings about foreign merchants whose goods threatened to upset the balance of trade. This xenophobia could become part of a wider distrust of foreign ideas, and as early as 1615 the economist Antoine de Montchrestien (1575–1621) had complained that 'Foreign doctrine poisons our minds and corrupts our morals'.[91] Although Fénelon seems to have shared the widely accepted view that the profit of Dutch merchants entailed France's loss, he nevertheless welcomed these traders, and his plans to restrict the French marine would have sustained this state of affairs. Monchrestien's xenophobic worries about the dangers of corruption by foreigners might have struck a chord in Fénelon's anti-luxury programme, but for the latter this possible risk was apparently outweighed by the perceived merits of the safety-valve function of surplus-taking foreign merchants. Moreover, Fénelon was more obsessed with the corrupting influence of goods than with foreign ideas, which, in his tightly regimented agrarian society, could be easily held in check anyway.

So, Fénelon's views on trade consisted of a preference for agriculture above commerce as the most vital economic sector, a criticism of money as an artificial source of wealth, rejection of a form of state intervention that aims to stimulate the balance of trade, a plea for 'free trade' (while at the same time endorsing restrictions on luxury trade), rejection of commercial autarchy, and benign views on Dutch merchants who are not only tolerated, but actually assigned a positive role in saving France from undue luxury by skimming off its surplus wealth. These views may seem disjointed, but actually they amounted to a consistent programme that accepted a substantial part of mercantilist economic and political *analysis* while completely rejecting and reversing its *policies*.

7 Quietism

It has been noted that Fénelon's complaints about social inequality, luxury, the misery of the common people, the disasters of war and its false glory fit into a long tradition of both secular and Christian moralism.[92] But it is possible to provide a more precise intellectual context for the three main concepts discussed here. Fénelon's aversion to luxury and war is closely related to a fear of chaos and change.[93] The quasi-totalitarian model offered by reformed

91 Montchrestien, *Traicté de l'oeconomie politique*, 92.
92 Cahen, introduction in Fénelon, *Télémaque*, vol. 1, xxv–xxvi.
93 Bonolas, 'Fénelon', 87.

Salentum seems designed to maintain a static society. In this model a nation of subsistence farmers is kept from mischief by constant toiling that can never produce luxury. The farmers are constantly supervised by a rigidly hierarchical elite that has been taught the dangers of offensive warfare. Limited trade acts as a safety-valve that siphons off economic surpluses so that luxury never gets a chance to trigger warfare and internal dissent. Since utopias tend to describe perfect societies, and since in a perfect society any change is for the worse, utopias often depict static societies. In that sense both Boetica and Salentum after the changes wrought by Mentor's reform programme are profoundly utopian.[94]

Fénelon's aversion to change and disorder in the outside world is matched by an equally stern verdict on the psychological state of turmoil caused by an undue hankering after luxury.[95] He is interested in both political and personal peace.[96] Both can be placed in the spiritual context of his quietism.[97] He had been drawn to the spiritualism of Madame Guyon since 1688 and amongst the many quietist works that he subsequently published, the *Explication des maximes des saints sur la vie intérieure* ('Explication of the maxims of the saints on the interior life') from 1697 ranks as one of the most important. The 'Explication' was Fénelon's own elucidation of the so-called 'Articles of Issy' in which Bishop Jacques-Bénigne Bossuet (1627–1704) and two other ecclesiastical dignitaries had stated the orthodox position of the Roman Church in regard to the ideas of Madam Guyon. Fénelon explains that we should aim for a pure or disinterested love.[98] This can only have the form of a love for God, considered in himself, without any added motive of hope or fear related to ourselves.[99] He discerns five degrees of love for God.

The fifth and highest degree of love takes the form of contemplation. The meaning of this concept can best be understood by a comparison with the fourth degree, which takes the form of meditation. In Fénelon's sense, meditation expresses the love of a person who certainly loves God, but not in a completely disinterested way. The meditator still retains a vestige of his own aims and desires, which means that he also remains in a position of hope and

94 See Trousson, *Voyages*, 100: 'c'est pour atteindre, en définitive, le même idéal de perfection statique, *ne varietur*, qui hante les utopistes'.

95 Bonolas, 'Fénelon', 85.

96 Lorson, 'Guerre', 211.

97 Cf. Mousnier, 'Idées Politiques', 206: 'Deux belles questions subsistent. De plus qualifiés chercheront les relations possibles entre la pensée politique de Fénelon et le quiétisme'; see also Niderst, 'Le Quiétisme', 205–16.

98 Le Brun, 'Notice' on Fénelon, *Explication des maximes*, i, 1534–9.

99 Fénelon, *Explication des maximes*, i, 1012.

fear. As long as he is wishing for some other reason than because it is God's will, he will remain troubled, and his views and tastes changing. He will be easily dissatisfied with others, disgruntled with himself, and full of defiance and distrust.[100] This less-than-perfect love is in constant need of two things. It needs to keep in constant view the many motives for its fears and hopes, and it also needs to assure itself that its various acts are in accordance with these motives.

Contemplation, on the other hand, is a kind of love that consists in simple, direct, and peaceful acts without any second thoughts.[101] Its sole purpose is to live in accordance with the will of God. While the love of God expressed by meditation still has something of an *esprit mercenaire*, persons engaged in contemplation are completely disinterested. They have the simple souls of children who are able to live in the present without caring for future needs.[102] Fénelon explains that contemplation does not imply a renunciation of the will; on the contrary, it implies willing one thing very much and that is to live according to God's will. This contemplation can be called passive in the sense that the soul is filled with peace and in the sense that it is not fuelled by spasmodic urge and unrest, but the soul is not without activity.[103] This consists in a series of constant and uniform acts that are expressed in constant and silent prayer. 'That is why', explains Fénelon, 'this contemplation has been called prayer of silence or *quiétude*'.[104] The childlike simplicity of quietist souls will prompt them to make a simple and peaceful use of all the things that surround them. In this they are completely at odds with what Fénelon calls the 'frantic licentiousness of the children of the [present] age'.[105]

It is not difficult to find Fénelon's spiritual values reflected in his political views as expressed in the *Telemachus* and in other works. The absence of luxury, war, and trade in simple and peaceful Boetica and the severe curtailment of these elements in more developed Salentum after Mentor's reform can even be read as a political implementation of Fénelon's quietism. We have already seen that he tried to provide a blueprint for a static society. Only a static society can guarantee the harmonious development of human relations and of virtue which is needed for true wealth, which is not of an exterior but of an interior nature.[106] Collective peace among people and nations and an inner peace of

100 Fénelon, *Instruction générale*, i, 655.
101 Fénelon, *Explication des maximes* i, 1060, 1088.
102 Fénelon, *Explication des maximes*, i, 1014–15.
103 Fénelon, *Explication des maximes*, i, 1081.
104 Fénelon, *Explication des maximes*, i, 1072.
105 Fénelon, *Explication des maximes*, i, 1078.
106 Bonolas, 'Fénelon', 89.

individuals, especially the inner peace of the future sovereign of France, are inextricably connected.[107] In the same way as spiritual contemplation of God supposes complete disinterestedness, a good king is disinterested in the sense that he does not desire his office for himself but only for the sake of his people. As Mentor explains to the Cretans: 'He that desires royalty does not know it: and how can he fulfil the duties of it, not knowing at all what it is? He desires it for his own sake; and you ought to desire a man who accepts it for your sake alone'.[108] A good king does not serve himself but only the laws of his realm.[109] This is the context for the frequent appearance of the word *désinterressé* in the *Telemachus*.[110]

Since 'disinterestedness' is clearly linked with simple contemplation and contrasted to discursive meditation, and since disinterestedness remains an essential concept after it has been transposed from a spiritual to a political context in the *Telemachus*, it is not surprising that an allusion to the distinction between contemplation and meditation can be found in the novel. This happens at the end of book IV, where Mentor has a theological (discursive) discussion on the coast of Cyprus with Hasraël about 'that supreme power that formed heaven and earth'.[111] This *discursive* discussion is only partly understood by Telemachus, who at that very moment is being regaled instead with a simple *contemplation* of the 'mild, yet majestic serenity' of the goddess Amphitrite.[112]

The passage about Amphitrite shows that Fénelon feels preciously little inhibitions in transposing key concepts of his Christian *spirituality* to an Ancient *political* context. In the essay *Sur le pur amour* ('On pure love') he took the opposite direction. Here he used a pre-Christian Ancient political context to make a spiritual point about the Christian faith. He points out that the idea of pure disinterestedness dominates the political theories of all ancient legislators.[113] Their political philosophy was based on a pure disinterested idea of friendship and citizenship. He compares this pagan ability to have a worthy motive for seeking a lower object (the city), with the tendency of many Christians to combine a worthy object (God) with the wrong motive, i.e. self-love.[114]

107 See Gilroy, 'Peace', 169.

108 Fénelon, *Telemachus*, 73/Fénelon, *Télémaque*, vol. 1, v, 229.

109 Fénelon, *Télémaque*, vol. 1, v, 190–1; see also Jaume, 'Fénelon critique', 395–422.

110 E.g. Fénelon, *Télémaque*, vol. 2, x, 61.

111 Fénelon, *Telemachus*, 55/Fénelon, *Télémaque*, vol. 1, iv, 171.

112 Fénelon, *Telemachus*, 56/Fénelon, *Télémaque*, vol. 1, iv, 177; see also Niderst, 'Le Quiétisme', 214.

113 Fénelon, *Sur le pur amour*, i, 665.

114 Riley, 'Rousseau', 87.

The relationship between spiritual and political disinterestedness is not only rich, ultimately it is also ambivalent. Although a spiritual contemplation of God is of a more elevated nature than a political life devoted to one's people, a king cannot lead a life devoted to contemplation. He has to act and if necessary, he has to wage war. A life of contemplation can only belong to a few 'souls that Heaven has preferred to separate from the others'.[115] If the king leads a life of political action rather than a life of spiritual contemplation, this can be regarded as the ultimate disinterested sacrifice in favour of his people. Spiritual values support the king to live his non-spiritual life.

8 Self-Interest

If Fénelon's interrelated views on luxury, war and commerce were informed by his spirituality in general and by his defence of *disinterestedness* in particular, then the question can be asked whether this notion was formulated in a reaction against forms of *self-interest*. Self-interest was indeed becoming an increasingly important early-modern notion, and various branches of learning, including philosophy and economics, seemed to co-operate in giving it a respectability that it had not enjoyed in previous ages.

The notion of self-interest suggests the satisfaction of desires and needs. In Aristotelian philosophy desires had been understood as the means towards the fulfilment of an end. The end of all men is happiness or *eudaimonia*—but once this aim has been reached, desire is supposed to be at an end, the mind at rest, and life complete. To pursue desires beyond their appropriate end would mean a lack of virtue. The notion of limited, instrumental, and objective desires was able to accommodate both a Classical concern with virtue and a Christian concern with salvation. Without these concerns happiness was not possible and both concerns implied a negative verdict on luxury, which merely facilitated the satisfaction of material needs with no contribution to virtue or salvation.[116]

Early modern mechanical philosophy can in many ways be understood as a long campaign against various forms of Aristotelian teleology.[117] The new philosophers had little sympathy for the curtailment of desires in the service of an objectively defined 'good life'. The resulting 'revaluation of desire', as it is dubbed by Christopher Berry, took many forms, and brought with it a

115 Fénelon, *Dialogues des morts*, i, 298; see also Niderst, 'Quietisme', 215; Jaume, 'Fénelon', 418.
116 Berry, *The Idea of Luxury*, 112–25.
117 See below, §3.5.

revaluation of self-interest.[118] Hobbes and Locke provide fine examples of this development. And in the previous chapter we have seen how De la Court's grand strategy is indeed informed by the rigorously self-interested rationality of his game-theoretic concepts.[119] In the present context I will pay special attention to John Locke (1632–1704), whose *Essay Concerning Human Understanding* was published in 1689, i.e. ten years before Fénelon's *Telemachus*. Locke's views on desire and self-interest are part of a psychology, an ethics, and a theology. In the *Essay* he explains that men are constantly beset by various kinds of uneasiness that are fuelled by a self-interested desire to increase their pleasure or happiness and a desire to decrease pain or misery. Desire is 'an *uneasiness* of the Mind for want of some absent good'.[120] Desires never come to an end, have no clear limits, and they cannot be defined objectively.

People tend to call pleasure good and pain bad. In itself this is a neutral anthropological observation, but there are moral implications. Locke never developed a complete ethics, but his hints in the *Essay* and in various manuscripts are clear enough. All our actions are either in conformity or disagreement with some law and in all cases the result will be pleasure or pain. In that respect there is no difference between moral laws and the laws of nature. I can commit murder and be punished, or I can put my hand on a hot stove and have it burned. In both cases I will be subjected to pain. Hence it does not come as a surprise that Locke uses the word 'good' for actions, for rewards, and for pleasure itself, and the word 'bad' for actions, punishments, and pain.

Although Locke's ethical Epicureanism does not allow him to make a fundamental distinction between moral laws and the laws of nature, he does make another distinction, which is again very Epicurean, namely that between long-term and short-term pleasures. In the fragment *Thus I thinke* (*c*. 1692), he gives a list of pleasures that include health, reputation, knowledge, and 'doeing good'.[121] While a good meal only gives a fleeting pleasure, 'the good turne I did (…) seven year since continues stil to please and delight me as often as I reflect on it'. The ultimate pleasure during our earthly existence consists in 'The expactacion of Eternall and incomprehensible happynesse in an other world'. This is again a clear instance of long-term pleasure over short-term pleasure. The inclusion of this item with the other items in the same list indicates that the moral rules of Christianity function along the same Epicurean

118 Berry, *The Idea of Luxury*, 112; see also Hont, *Jealousy of Trade*, 42; Jennings, 'Luxury', 82; Wahnbeck, *Luxury*, 20.

119 See above, §1.4.

120 Locke, *An Essay*, II.xxi.31, 251.

121 John Locke, *Thus I think*, 1.

lines as other moral rules. In all cases, pleasure and pain are the moral instruments *par excellence*. In *Of Ethic in General* Locke remarks that goodness would signify nothing 'were there noe losse of pleasure, noe paine to follow from a mans satisfying his appetite as he could'.[122] So Locke's psychology, his ethics, and even his theology are built around self-interested hedonistic individuals. Although he has no doubt about the universal and objective value of God's laws, in the fragment *Ethica A* he points out that every ethical system should take due consideration of the subjective needs of each individual, or 'each mans particular disease', which is formed by 'the pleasure that possesses him'.[123] Finally, although Locke does not say much about luxury, it follows from his Epicureanism that this is not necessarily the object of a moral verdict, and that it is reprehensible only insofar as it will tend to favour short-term pleasures over long-term pleasures.

While mechanical philosophers tended to analyse the existence of self-interest on an individual level, and derive a certain satisfaction from a subsequent step in which they would tend to restrict its moral implications, economic thinkers developed their own way of neutralizing the spectre of self-interest. Their point of departure was a separation between luxury in a moral context applied to *individuals* and luxury in a political context applied to the *state*. To this they added a notion of well-being that was defined in terms of economic prosperity. They believed that moral condemnation of self-interested individual luxury should not necessarily be extended to 'political' luxury, since luxury could contribute to the material well-being of the nation. From this it was only a small step to a depreciation of both the Classical and the Christian concern with virtue, since neither was well suited to accommodate the materialistic elements of this new form of well-being.[124]

A good example of the economic strategy to de-moralize luxury is provided by the English economic thinker Nicholas Barbon (1640–1699) in his *Discourse of Trade* (1690). He makes a distinction between an individual's 'wants of the body', concerning food, shelter, and clothing, and so-called 'wants of the mind' for all additional goods, i.e. luxury goods. While the wants of the body are limited and objective, those of the mind are infinite and subjective. Yet in spite of their subjective character these wants are 'as natural to the Soul, as Hunger to the Body'.[125] Barbon's use of a 'mind'–'body' terminology to make a distinction between a desire for necessary and for luxury goods is slightly misleading, as

122 Locke, *Of Ethic in General*, fol. 149v.

123 Locke, *Ethica A*, 224.

124 See Berry, *Luxury*, 101, 125; Wahnbeck, *Luxury*, 14, 16–17.

125 Barbon, *A Discourse of Trade*, 15; see also Kellow, 'Strength and riches', 1–22.

the 'wants of the mind' can also pertain to our body, as transpires from the following quotation: 'Amongst the great Variety of things to satisfie the Wants of Mind, those that adorn Mans Body, and advance the Pomp of Life have the most general Use'.[126] A second distinction made by Barbon is that between persons and nations. While a person's resources are finite, those of a nation are infinite; its animals 'naturally increase', its agricultural produce is renewed every year, and the supply of its minerals is inexhaustible.[127] So, while there may be good sense in curbing luxury expenditures in a person, there is no reason at all to impose frugality and sumptuary laws on an entire national economy.

Actually, the procurement of luxury goods by rich individuals aids the poor and stimulates trade, which in its turn raises national prosperity. A good example of this mechanism, according to Barbon, was France, where Louis XIV had stimulated a rivalry among the members of his aristocracy that was expressed by a conspicuous consumption. Taxation of this consumption had done a lot to increase the King's revenues.[128] Barbon points out that trade also plays a crucial role in relation to war. In previous times, trade had been considered prejudicial to the growth of empires, because it tended to soften people by 'Ease and Luxury', which made them unfit to endure the rigours of warfare.[129] In the seventeenth century, the author notes, this has changed completely. Trade has become necessary to provide weapons, it 'Provides the Magazines of Warr. The Guns, Powder, and Bullets, are all made of Minerals, and are wrought by *Traders*'.[130] Thus trade boosts the military capabilities of a country, as is shown again by the case of Louis XIV, who has 'become troublesome to his Neighbours'.[131] But this is only one side of trade's complicated relationship with war. Trade has also a potential to further peace. When the barbarian tribes that had invaded the more civilized parts of Europe for centuries finally became 'settled in *Trade*', the wines and spices from more civilized parts of the continent made their own regions more habitable, which took away much of their aggressive fervour. Moreover, the growth of commerce in barbarian provinces increased wages at home, making it less attractive to indulge in plundering abroad.[132] Hence self-interested and unlimited desire results in luxury, but whereas luxury on a private level may end in bankruptcy, luxury on the

126 Barbon, *A Discourse of Trade*, 15.
127 Barbon, *A Discourse of Trade*, 5–6.
128 Barbon, *A Discourse of Trade*, 69.
129 Barbon, preface in *A Discourse of Trade*, [iv].
130 Barbon, *A Discourse of Trade*, 40.
131 Barbon, *A Discourse of Trade*, 70.
132 Barbon, *A Discourse of Trade*, 37–9.

aggregate level of the state stimulates trade, which in its turn stimulates the military strength of a nation. Thus Barbon prefigures Bernard Mandeville's (1670–1733) famous dictum that 'Private Vices by the dextrous Management of a skilful Politician may be turn'd into Publick Benefits'.[133]

So, the growing prosperity in the seventeenth-century Dutch Republic and other parts of north-western Europe, prompted various reactions. The process had its apologists but it also left growing resentment and attacks on luxury, self-interestedness, and related issues in its wake.[134] In the seventeenth-century, defenders of the process tended to be most vociferous in England, while criticism was perhaps most articulate in France. The word *luxe* entered the French language in 1606 as a synonym for 'superfluity', but by 1694 its meaning in the *Dictionnaire de l'Académie Française* had shifted to a sterner 'excess', which implied a clear note of moral condemnation.[135] In the case of the Duke de La Rochefoucauld (1613–1680), criticism of selfish behaviour took the form of a subtle *démasqué*. The traditional aristocratic code of conduct had prescribed that generosity and gratitude should be 'unselfish'. Often the point of the Duke's maxims was that *actually* an exchange was taking place that was far from altruistic. In the words of Pierre Force, 'aristocratic disinterestedness is the customary way of pursuing aristocratic interests'.[136] Less subtle, in his *Moeurs des Israélites* (1681), the Abbé Claude Fleury (1640–1723) denounced the pernicious consequences for society of luxury consumption stimulated by the mercantilist system.[137] Finally, Blaise Pascal (1623–1662) developed an ethics in which the importance of self-denial was formulated in an express reaction against self-interest. Indeed, in the words of Pierre Force, 'A paradoxical consequence of the interest doctrine and of the critique of the interest doctrine is the emergence of disinterestedness as a moral value'.[138]

This is the wider intellectual background for Fénelon's defence of disinterested pure love of God, the attack on luxury and war, and the heavily circumscribed role assigned to trade. His defence of *disinterestedness* in both (a) a spiritual and (b) an economic and political context is the precise opposite of (a) a defence of *self-interest* on an ethical and theological level, as for example in Locke, and (b) a defence on an economic and political level, as for example

133 Mandeville, *The Fable*, 371.

134 A similar process had been seen early in Renaissance Venice and in Rome during the late Republic and the Empire. See Sekora, *Luxury*, 71.

135 Jennings, 'Debate', 82.

136 Force, *Self-interest*, 181.

137 Kapp, *Télémaque de Fénelon*,114.

138 Force, *Self-interest*, 183.

in Barbon. Both Locke and Fénelon present a psychology that starts with the feeling of uneasiness. This is used by Locke as the point of departure for a psychology that is built around mechanical stimuli, while Fénelon advocates a state of *quiétude* by constant prayer to God. According to Locke, uneasiness fuels a constant quest for happiness and this is seen as a legitimate pursuit. Fénelon, on the other hand, is interested in virtue and salvation, not in personal happiness, as his grim blueprint for Salentum makes abundantly clear. He has an uncanny talent for detecting and rejecting self-interest in our moral conduct and in our love for God. Only the fifth degree of love is completely disinterested. The first and lowest degree of love for God consists of what he describes rather nastily as the 'love of the carnal Jews for the gifts of God, as distinguished from himself'.[139] In many ways this is exactly the kind of love that is expressed, with a refreshing lack of any scruples, by Locke's Christian Epicureanism.

While Barbon distinguishes a private and a public sphere, where self-interest on the former level results in benefits on the latter level, there is no such separation in Fénelon, whose political philosophy can be read as a projection of quietist notions of individual peace and simplicity. Barbon and Fénelon nevertheless share to a surprisingly high degree the same analysis of a connection between luxury and trade. Since Barbon does not reject military spending, he leaves the door wide open for a self-reinforcing mechanism consisting of luxury, trade, and war. In Fénelon's political philosophy, on the other hand, everything is geared towards a destruction of this mechanism.

9 Evaluation

Fénelon dared to challenge the luxury of the court of Versailles and the crippling tax burdens for the peasants, and it has been suggested that the National Assembly echoed Fénelon when it passed a motion that denounced 'war waged for the honour and glory of kings'.[140] More recently, Leslie Shaw has noted that Fénelon's ideas 'are surprisingly close to the modern concepts of sustainable development, ecology, urban-rural balance, corporate responsibility and ethical trade'.[141] But Fénelon was by no means a prophet of modernity; he aimed to reform his country and then freeze it into the structure of a static agricultural society, at a time when Europe was fast becoming the hub of a dynamic global

139 Fénelon, *Explication des maximes*, i, 1012.
140 Bell, *The First Total War*, 108.
141 Shaw, 'Fénelon'.

commercial network.[142] While Fénelon's stress on agriculture in favour of commerce has been described as pre-physiocratic, he had none of the liberal aversion to state intervention that marked most eighteenth-century physiocrats, and his own blueprint for a reformed France had totalitarian implications that went far beyond the crassest forms of Bourbon despotism.[143] Moreover, while Fénelon's admiration for simplicity, martial virtue and beauty was combined with a pacifist agenda, many of his eighteenth-century followers would idealize primitive and Spartan virtues without similar compunctions.[144] Finally, the implementation of Fénelon's ideas by his royal pupil would have resulted in de-urbanization, swept away France's luxury trade, damaged its commerce and hence crippled the financial basis for its political and military status as a great power, and would have established a police state at home in the territories not overrun by its enemies—all in the name of peace and simplicity. Insofar as Enlightened philosophy was concerned with political freedom and socio-economic development, Fénelon's political quietism is difficult to surpass as its very counter-image.

142 Cf. Mousnier, 'Fénelon', 194 and Racault, *Utopie*, 202; see also Rothkrug, *Opposition*, 297–8: 'Fénelon was both the last of the great seventeenth century mystics and a leader of the first large-scale encounter between those who were convinced that moral progress was possible and those who denied that human efforts could significantly alter man's fundamentally corrupt nature. At bottom, this was the issue between the movement for reform and the forces of tradition'.

143 Racault, *Utopie*, 201; Bonolas, 'Fénelon', 89.

144 Paret, *Modern Strategy*, 107–8; see also Regent, 'In the Shadow', 477–509.

CHAPTER 3

'The Effect in Turn Became the Cause': Determinism and Causal Feedback Loops in Montesquieu's Explanations for the Military Rise and Fall of Rome

1 Introduction

The most famous work by the political philosopher Charles-Louis de Secondat, Baron de La Brède et de Montesquieu (1689–1755) is his monumental *De l'Esprit des lois*, published in 1748 (*The Spirit of the Laws*). In the present chapter I will focus on a much smaller and less well-known essay: the *Considérations sur les causes de la grandeur des Romains et de leur décadence*, 1733/1734 (*Considerations on the Causes of the Greatness of the Romans and Their Decline*).[1] The *Considerations* are very much an analysis of the causes of military strength and weakness. Richard Myers remarks that the importance of warfare in the *Considerations* was missed by most commentators, but it is emphatically present in Roger Oake's terse summary: 'One should note the militarist chain: Rome's "principle of continuous warfare" leads to the expansion of her frontiers; she pursues a foreign policy which is entirely the tool of a general staff; she is forced to maintain large armies in the field, to expand and control the empire thus obtained; the citizen-armies, kept long in absence from home, lose their character and become efficient professional tools of their military leaders; to keep up the strength of these armies, such tremendous numbers of foreigners must be admitted to citizenship that the homogeneity of the state is destroyed'.[2] Montesquieu's interest in the causes of Rome's military rise and fall is part of a more general interest in what exactly makes nations strong or weak. In this essay I will focus on the formal and conceptual nature of Montesquieu's explanations. I am especially interested in the precise role of determinism versus contingency and personal

1 Montestequieu, *Considérations sur les causes de la grandeur des Romains et de leur décadence* in *Oeuvres complètes de Montesquieu*, vol. 2, 87–318, ed. Françoise Weil and Cecil Courtney (Oxford: Voltaire Foundation, 2000); translation: Montesquieu, *Considerations on the Causes of the Greatness of the Romans and Their Decline*, transl. David Lowenthal (Indianapolis: Hackett, 1999).

2 Oake, 'Montesquieu', 53; Myers, 'Montesquieu', 43–4; cf. Larrère and Weil, introduction in Montesquieu, *Réflexions*, ed. Weil, 32, 332; Strosetzki, 'Dekadenz', 78.

 | DOI:10.1163/9789004536678_005

agency in the *Considerations*. I will try to gain a deeper understanding of this issue through an analysis of Montesquieu's method of process explanations in general and his use of causal feedback loops in particular. The result will be placed in two wider contexts. The first context is military and political. The writing of the *Considerations* was influenced by the recent reverses of France during the War of the Spanish Succession (1701–1714).[3] The second context is philosophical and historiographical. I will argue that Montesquieu used and adapted four important elements of early modern mechanical philosophy for the development of a modern historiography that focuses on dynamic processes rather than static substances, is partly reductionist but also partly holistic, sees history as partly contingent and partly determined, and shows a radical commitment to explanations in terms of efficient causality rather than final causality.

2 Determinism and Contingency at the Meso-level

The *Considerations* is not a history textbook, although it would be used as such until the twentieth century.[4] Nor is it a work of conventional military history, with a focus on separate events and separate decisions of generals. The *Considerations* can rather be read as a methodological exercise in causal explanation applied to the subject of the military rise and fall of Rome.[5] Montesquieu's treatise on Rome is situated on the explanatory meso-level, i.e. the level of separate civilizations, empires or nations. This level can be located between the level of micro-explanations of individual actions and events, and the macro-level of universal laws of human history in general.[6] This meso-level is where Montesquieu liked to make comparisons. Thanks to the *Spirit of the Laws* he has laid claim to the honour of being the first comparative political sociologist; but already the *Considerations* abounds with comparisons between Rome and previous, contemporary and later polities.[7] A good example of a meso-explanation that involves a comparison of two different polities

3 Rahe, 'The Book', 26; Rahe, *Montesquieu*, 6–26; see also above, §2.2.
4 Bomel-Rainelli, 'Les rapports', 146; Andrivet, introduction in Montesquieu, *Considérations*, 14.
5 Lowenthal, 'The design', 161.
6 Cf. Little, 'Change', 89–91.
7 See also the preface to Montesquieu, *Esprit*, 229: 'Quand j'ai été rappelé à l'Antiquité, j'ai cherché à en prendre l'esprit, pour ne pas regarder comme semblables des cas réellement différents, et ne pas manquer les différences de ceux qui paroissent semblables'; Aron, *Les étapes*, 65; Benrekassa, *La politique*, 61; Lottes, 'Montesquieu', 115; Courtois, *Inflexions*, 29.

is Montesquieu's answer to the question why the Second Punic War was won by Rome and not by Carthage, included in the long 'Comparison of Carthage and Rome' in chapter 4 of the *Considerations*. While Rome's modest wealth was distributed relatively evenly, Carthage's enormous commercial riches were reserved for the privileged. While the Roman legions consisted of well-motivated citizens who felt they had an equal stake in their Republic, Carthage used its wealth to hire mercenaries who were prone to mutiny. While the simple virtues of the Romans proved an endless source of strength, Carthaginian wealth exhausted itself. And while Rome was united in the commitment to its wars, the Carthaginians were constantly divided between a war faction and a peace faction.[8]

In historiographical thought there seems to exist a certain correlation, by no means rigid and mechanical, and often accepted implicitly rather than formulated explicitly, between the level of generality of historical explanations and their commitment to historical determinism.[9] By historical determinism I mean the view that historical events are predetermined and inevitable, and in that sense *necessary*, while the opposite view holds that historical events are the result of contingent facts and free decisions. Narrative interpretations of singular events on a micro-level tend to stress historical contingency, individual agency, and the multiplicity of possible outcomes. Explanations on a macro-level, on the other hand, tend to focus on the determinist force of broad structures and processes. Hence it is not surprising that explanations on the meso-level tend to embrace a mixture of determinist and non-determinist elements. Daniel Little describes these explanations at the 'conjunctural contingent meso-level' as 'explanations that identify intermediate-level structures and processes and highlight both the structural factors that govern change and the multiple pathways that change can take'.[10]

There is reason to believe that Montesquieu entertained similar views on a correlation between the level of generality and the degree of determinism. In the short *Essai sur les causes qui peuvent affecter les esprits et les caractères*, produced between 1736 and 1741 ('Essay on the causes that can influence minds and characters'), he wrote that 'these causes/the physical causes become less arbitrary to the degree that they have a more general effect'.[11] In the specific

8 Montesquieu, *Considérations*, iv, 110–15.

9 Cf. Aron, *Les étapes*, 61.

10 Little, 'Change', 89–91.

11 Quoted in Courtois, *Inflexions*, 27: 'ces causes/les causes physiques deviennent moins arbitraires à mesure qu'elles ont un effet plus général'.

case of his *Considerations*, most historians have preferred a determinist interpretation of this 'history without heroes'.[12] Indeed, the work contains many passages that support such a view. In the following quotation Montesquieu makes a general statement about historical determinism—note his use of the term 'necessary': 'There are general causes, moral and physical, which act in every monarchy, elevating it, maintaining it, or hurling it to the ground. All accidents are controlled by these causes. And if the chance of one battle—that is, a particular cause—has brought a state to ruin, some general cause made it *necessary* for that state to perish from a single battle. In a word, the main trend draws with it all particular accidents'.[13] There are many examples where we see the same commitment to determinist explanations. In the early Roman Republic wealth was despised, poverty was regarded as a virtue, and neither birth nor personal dignity allowed the patrician class to accumulate a lot of power. Hence power was *necessarily* shared by the largest number of citizens.[14] During the civil wars of the late Republic, rich citizens who were in favour of traditional liberty and who declined to take the side of any of the generals who were tearing apart the fabric of the Republic, ran an increasing risk of losing their money in the ever more frequent proscriptions. In order to protect their wealth, they felt forced to take sides. Hence the number of citizens who defended liberty dwindled and hence the Republic had to perish *necessarily*.[15] Once Augustus had won the Civil War there was no way back for Rome, and even if he had not wanted to, a form of government controlled by one single person had become *inevitable*.[16] In a similar vein, Montesquieu notes that if Cyrus had not conquered the kingdom of Lydia, Persia subsequently would never have been conquered by the Macedonians. Similarly, if Seleucus had remained in Babylon and left Syria to the Antigonids in the west, his empire would not have been annexed by the Romans. Montesquieu uses both particular observations to make the more general point that provinces at the periphery separated by geographic barriers from the central power base of an empire, such as the mountains of Asia Minor and the Syro-Arabian Desert, are difficult to control in the long term. This in its turn leads to the even more

12 Ehrard, 'Rome', 30: 'une histoire romaine sans héros'; Goyard-Fabre, 'Rome', 124–9; Goyard-Fabre, *Montesquieu*, 29, 52–3; see also Aron, *Les étapes*, 37.

13 Montesquieu, *Considerations*, 169/Montesquieu, *Considérations*, xviii, 235 (italicization by me PS).

14 Montesquieu, *Considérations*, viii, 145.

15 Montesquieu, *Considérations*, xi, 165.

16 Montesquieu, *Considérations*, xiii, 188.

general determinist observation that 'Nature has given states certain limits to mortify the ambition of men'.[17]

Although the determinist import of the *Considerations* is undeniable, Robert Shackleton remarks that according to Montesquieu the destiny of Rome was governed by a multitude of causes, and 'Of these causes some were necessary and others were accidental'.[18] I agree with this verdict, but I think that Montesquieu's choices between necessity and contingency (or personal agency) followed a distinct pattern that can be understood against the background of his process explanations.

3 Process Explanations

Ever since Plato and Aristotle, philosophers have favoured accounts of reality in terms of timeless and permanent substances over explanations involving dynamic processes. Eighteenth-century interest in history, however, went hand-in-hand with a renewed theoretical appreciation of processes of change. Giambattista Vico (1668–1744) maintained that the *nature of things* can be sufficiently explained by an account of *how they come into being*.[19] Montesquieu shared Vico's interest in historical change, but he deployed his own conceptual apparatus.[20] He considered Rome as a whole, as a system, in the broad sense of an organized and connected group of entities, consisting of various institutions, customs, and mentalities. Rome was a *Système total*, a system that formed one *grand Corps* of which all the parts were interconnected.[21] Once such a system has been established and once things have been arranged in a certain way, it is usually best to leave it unchanged. The reasons that have kept a political system going in the past or often the same reasons that keep it going in the presence. These reasons are extremely complicated and they are largely unknown. If one tries to change the whole system, only part of the problems can be resolved in advance ('in theory'), while other problems will have to be dealt with later ('in practice'). For example, according to Montesquieu, the late Roman Empire was ruined by the decision to divide it in a western and an

17 Montesquieu, *Considerations*, 61/Montesquieu, *Considérations*, v, 126.

18 Shackleton, *Montesquieu*, 166; see also Lottes, 'Montesquieu', 112–3.

19 Perinetti, 'Philosophical reflection', 1127–8.

20 Cf. Aron, *Les étapes*, 65; Goyard-Fabre, *Montesquieu*, 1; Andrivet, introduction in Montesquieu, *Considérations*, 30.

21 Montesquieu, *Considérations*, xvii, 228; see also Vuia, *Montesquieu*, 47; Senarclens, 'Herangehensweise', 131.

eastern part. The parts of the complete system continued to depend on each other, even after this system had been torn apart.[22]

Montesquieu described the initial state of the Roman system and he then gave an account of the process by which it evolved into later states. He provided process explanations for the change of a historical system.[23] Historical systems can be more or less determined, and process explanations are not inherently determinist. They are merely committed to the claim that a specific process can bring about the explanandum in question and that this is indeed what happened—not that this happened *necessarily*.[24] Nevertheless, Montesquieu's process explanations evoke a highly determined system that was active throughout Rome's long history. At the same time, Montesquieu notes that the initial state of the Roman system at the birth of the Republic was highly *contingent and undetermined*. In that particular situation, relatively small and insignificant events and decisions were sufficient to change the course of Roman history. The rape of Lucretia by Sextus Tarquinius, son of King Tarquinius Superbus, sufficed to produce the fall of the monarchy in 509 BC. Several factors, each of them again relatively contingent, assured that warfare (or more precisely territorial expansion) became the main principle of the new republic.[25] Since the consuls were elected for one year only, they were motivated to realize their ambitions swiftly. The presence of two consuls institutionalized a constant rivalry that acted as another stimulus to military initiatives. Moreover, the Senate was constantly harassed by the demands and complaints of the people, and tried to deflect this political threat by turning it outwards to military campaigns. The people were poor, Rome was without commerce, and the prospect of military loot was most welcome. Military discipline was maintained by the holiness of an oath that gave each citizen-soldier an equal share in the spoils of war. These factors all favoured the development of the martial virtues of constancy and valour. The intimate link between the origins of the Roman Republic and its warlike character are reflected in the dual title of the first chapter of the *Considerations*: '1. Beginnings of Rome 2. Its Wars'.[26] The Romans dedicated themselves to warfare; this became their sole art and they put 'their whole spirit and all their thoughts into perfecting it'.[27]

22 Montesquieu, *Considérations*, xvii, 228.

23 Botterill, 'Causal Explanation', 388; Rueger, 'Connection', 78–79.

24 Botterill, 'Two Kinds', 305.

25 On *principes* see Montesquieu, *Considérations*, i, 94; vi, 135; xiii, 189. For the difference between the term *principe* in the *Considérations* and the *Esprit* see Myers, 'Montesquieu', 42–3.

26 Montesquieu, *Considerations*, 23/Montesquieu, *Considérations*, i, 89.

27 Montesquieu, *Considerations*, 33/Montesquieu, *Considérations*, i, 99.

Rome was 'made' for military expansion and its laws and institutions were admirably tailored towards this objective.[28]

Montesquieu observes that although the initial state of the Roman system was very contingent, its warlike institutions and mentality would prove extremely resilient in the process that followed. The same principles used against the small villages at the beginning of Rome's history were still used successfully much later against the most powerful monarchies.[29] During its expansion after the defeat of Carthage, Rome would become increasingly corrupt. But such was the strength of its militarist mentality, that it would survive long after the circumstances by which it had been favoured had vanished: 'Yet, whatever the corruption of Rome, not every misfortune was introduced there. For the strength of its institutions had been such that it preserved its heroic valour and all of its application to war in the midst of riches, indolence and sensual pleasures—which, I believe, has happened to no other nation in the world'.[30] At the start of the Principate 'the martial virtues remained after all the others were lost'.[31] When the all-pervading moral corruption began to destroy Roman military art as well, when soldiers could no longer be motivated to carry the heavy weight of their arms and armour, and when they stopped fortifying their camps, the Empire would prove defenceless against the onslaught of the barbarians.[32]

In the sequence described here, there is a clear difference between the initial state of the system and the subsequent stages of the process. The process itself is highly determined, and thus accords very well with the kind of determinism that most commentators have ascribed to the *Considerations*. But the process is constrained by an initial state that consists of highly contingent facts and voluntarist actions. In this initial state the impact of individual actions on institutions and mentalities is large, but in the subsequent stages it is small. Or, in the words of Montesquieu: 'At the birth of societies, the leaders of republics create the institutions; thereafter, it is the institutions that form the leaders of republics'.[33]

28 Montesquieu, *Considérations*, ix, 158.

29 Montesquieu, *Considérations*, vi, 135; see also *Esprit* viii,13, 360: 'Il n'y a point eu de peuple, dit Tite-Live, où la dissolution se soit plus tard introduite que chez les Romains, et où la modération et la pauvreté aient été plus longtemps honorées'.

30 Montesquieu, *Considerations*, 98/Montesquieu, *Considérations*, x, 162.

31 Montesquieu, *Considerations*, 99/Montesquieu, *Considérations*, vi, 163.

32 Montesquieu, *Considérations*, xviii, 236–7.

33 Montesquieu, *Considerations*, 25/Montesquieu, *Considérations*, i, 90 (in edition of 1758).

The phenomenon that Montesquieu here very perceptively observes, is still an object of research in the social sciences and has been dubbed 'institutional momentum' by Daniel Little, who remarks that 'institutional configuration is plastic in its development and relatively sticky in operation'.[34] In the course of history, groups of people have been confronted with certain challenges to improve their conditions of life. Similar material circumstances allowed for different rational choices and hence different institutional solutions. Moreover, even with purposive actions by separate individuals, the institutional effects could still be completely unintended. Once these institutional solutions were in place, they started to have their own momentum. In this way contingent beginnings were followed by determined consequences.[35] Little places these contrary tendencies very much on the level of meso-history, where we can 'discover both the unifying dynamics and the differentiating expressions which these abstract processes take in different historical settings'.[36]

But how exactly *do* institutional solutions gain momentum? Or, more to the point, how could the *plastic* initial configuration of the Roman system become so *sticky* in its subsequent operation? This is not be a trivial question for a modern historian and it certainly was not a trivial question in the eighteenth century. Yet Montesquieu appreciated the question and he provided a deep and comprehensive explanation for the plastic/sticky sequence. This explanation can be understood against the background of what I consider to be at the heart of his process explanations, i.e. his use of feedback mechanisms.

In a feedback mechanism, part of the output influences the input. In the case of positive feedback, the output and the input go in the same direction, i.e. both increase or both decrease. If both output and input *increase*, the system will spiral upward towards an 'explosion'. An example of this is an escalating arms race. If both output and input *decrease*, the system will spiral downward towards an 'implosion'. An example of this is the process of de-escalation after two opponents have reached an agreement on weapons reduction. In both cases output and input have the *same* direction. The mechanism is self-reinforcing, and the feedback movement will be repeated an indefinite although not an infinite number of times. If output and input go in a *different* direction, however, the result will not be self-reinforcing but balancing; if part of an increasing output is used to decrease the input, the result will be a state of equilibrium before either an explosion or an implosion has occurred. The same equilibrating effect occurs in the opposite situation, i.e. a situation

34 Little, 'Change', 108.
35 Little, 'Change', 106–9.
36 Little, 'Change', 109.

when part of a decreasing output is used to increase the input. These kinds of balancing feedbacks are called negative.[37] An example of negative feedback is a heating system with a thermostat. Increased temperature will prompt the thermostat to switch off the heating system, resulting in decreased temperature; which, after a certain delay, will prompt the thermostat to switch on the heating system again. While positive feedback loops by their exponential character tend to change or even destroy the system, negative feedback loops tend to stabilize and preserve the system. They tend to keep the variations in a system inside a certain bandwidth. Feedback loops can be regarded as general mechanisms that underlie every historical empire or system. At the same time, the precise character, combination, and potency of the feedback mechanisms in each system will vary.

Although feedback processes seem well suited to Montesquieu's explanations for the military trajectory of Rome's history and the military mentality of the Romans themselves, the question is of course whether Montesquieu was at all acquainted with these concepts and whether indeed he used them to this effect in the *Considerations*. The answer to both questions is affirmative. The 'militarist chain' that Oake discerns in Montesquieu *Considerations* (see the Introduction) had the form of feedback processes. I will concentrate on Montesquieu's use of the *concept* of feedback processes; but in one case he even comes very close to its *terminology*. When he describes the instability in the Byzantine Empire in the sixth and seventh centuries, he gives an exact description of a feedback loop, one in which instability breeds even more instability (italics added by me): 'The revolutions that occurred themselves gave rise to other revolutions, and *the effect in turn became the cause*'.[38] Let us now see how Montesquieu's account of the rise and fall of Rome consists of a dynamic combination of different feedback loops.

---

37 See Golec, 'Feedback'; Richardson, 'Systems and complexity', 76–7; Jervis, 'Complex Systems', 20.

38 Montesquieu, *Considerations*, 198/Montesquieu, *Considérations*, xxi, 262. The concepts and terms of positive and negative feedback were articulated in a formal way by Ludwig von Bertalanffy in his general system theory from the 1930s onwards (Bertalanffy, foreword in *General System Theory*, vii–ix). However, the concept of *feedback*, and more specifically the concept of *negative* feedback (which is a bit more complicated than *positive* feedback) was already understood and applied by Ktesibios of Alexandria (285–228 BC), who used it for a self-regulating water-clock. The Dutch inventor Cornelis Jacobsz Drebbel (1572–1633) applied the same concept to the first prototype of the thermostat. Negative feedback was also used for the automatic control of windmills in eighteenth-century Britain (Mayer, 'The Origins'; Snelder, 'Drebbel', 273–5).

Montesquieu reports how at the beginning of its history, Rome was surrounded by various tribes that were all very warlike. In this tough competitive environment Rome necessarily had to either perish or prevail.[39] Given the nature of the particular feedback loops that determined the Roman system, its triumph over all its neighbours was inevitable.[40] The harsh military environment of Latium provided a good training ground for martial skills. With each defeat Rome became only more determined to win. Such was its mentality and such were its institutions, that it never ended a war that it did not win.[41] Whilst engaged in this process, Rome 'within a very small orbit, practised the virtues which were to be so fatal to the world'.[42] The principle of warfare and a belligerent mentality or *esprit général* reinforced each other in a feedback loop. This loop was repeated over a long period and it became engrained in the Roman system. The self-reinforcing character of the process solidified contingent beginnings. This explains how Rome's plastic initial configuration became so sticky in its subsequent operation.

This phase in Rome's history lasted from the birth of the Republic until the victorious termination of the Second Punic War. Even the shattering defeats at the hands of Hannibal only served to augment the resolve of Rome: 'It is true that at first the terror in Rome was extreme, but the consternation of a warlike people, which almost always turns into courage, is different from that of a vile populace, which senses only its weakness'.[43] Once Rome had defeated Carthage, a second feedback loop began to dominate the history of the Republic: 'Rome continually grew richer, and every war put it in a position to undertake another'.[44] While the first feedback loop had shaped Rome's military mentality in adversity, the second loop, consisting in war feeding on itself, furthered Rome's subsequent expansion. Military success and the wealth this gained allowed the annexation of the Hellenistic monarchies and catapulted Rome to universal supremacy. Hardened by its relatively small successes in the great wars against Carthage, Rome was in a position to profit from great successes in the relatively small wars in the Eastern Mediterranean.[45] Once Rome's enemies were defeated, their finances were ruined to Rome's profit. Actually this inexorable process was not only fuelled by Rome's enemies, but

39 Montesquieu, *Considérations*, i, 94.
40 Montesquieu, *Considérations*, i, 94.
41 Montesquieu, *Considérations*, i, 91–4.
42 Montesquieu, *Considerations*, 29/Montesquieu, *Considérations*, i, 97.
43 Montesquieu, *Considerations*, 51/Montesquieu, *Considérations*, iv, 118.
44 Montesquieu, *Considerations*, 73/Montesquieu, *Considérations*, vi, 138–9.
45 Montesquieu, *Considérations*, v, 121.

also by its friends. Rome's allies wrecked themselves to present Rome with gifts that were intended to preserve a peaceful relationship, thus providing Rome with the means to augment its supremacy even further, while only half this money would have sufficed to pay for an army that would have been capable of defeating Rome.[46]

Rome's military expansion would eventually be checked and Montesquieu describes several feedback loops that were responsible for this process. The core of Rome's military power had always consisted of its citizen armies. As the Republic expanded, however, its soldiers remained far from home for increasingly longer periods and started to transfer their loyalty from the Republic to the individual generals with whom they were on campaign. Loot and conquest allowed these generals to pay their soldiers, which alienated these armies even more from the Republic.[47] While the equal partition of land among citizen-soldiers had been a key economic institution of the Republic, Sulla and his successors would take the distribution of farms to veteran soldiers into their own hands.[48] As the Republic expanded and hence offered an increasing number of geographical power bases, professional armies whose first loyalty was to political generals multiplied. These generals were bribed by Rome's allies and friends, which enabled them to recruit even more professional soldiers. Moreover, as Rome expanded, citizenship was granted to an ever-widening group of foreign people who could not identify with the old institutions and whose vote could easily be bought by magistrates who had enriched themselves in the provinces of the expanding Republic.[49] These military and political developments severely eroded Rome's political system, and facilitated civil wars that eroded the system even further. So these feedback loops explain how Rome's territorial rise and increasing wealth had consequences that would slowly but inescapably cause its fall.

The inevitable process of dissociation between Rome's soldiers and its citizens became more pronounced during the Empire, when the legions at the borders allowed the people to live under the *Pax Romana* inaugurated by Augustus. Although it had been very resilient, Rome's military mentality was eventually affected; while warfare had been the principle of the Republic, peace increasingly became the principle of the Empire. This change spelled doom for the Romans, 'who had a continuous sequence of successes when they were guided by a certain plan, and an uninterrupted sequence of reverses

46 Montesquieu, *Considérations*, vi, 130–41.
47 Montesquieu, *Considérations*, ix, 153.
48 Montesquieu, *Considérations*, xi, 165.
49 Montesquieu, *Considérations*, ix, 154–9.

when they followed another'.[50] Victories became a source of concern to the emperors, since victorious armies asked higher remunerations and threatened the authority of their emperor. So in order not to provoke the jealousy of their emperor, generals learnt to moderate their glory.[51] This weakened the military mentality of the Empire even more. The Empire started to buy off the barbarians to the point where the transfer of wealth from the Empire to the barbarians significantly altered the power relations between them.[52] In this way the Empire cycled inexorably downwards to its eventual dissolution.

Montesquieu not only had a keen appreciation of the various feedback loops that contributed to Rome's growth as a world power and its eventual decline. He also showed a striking, yet rigid, grasp of the relation between the shape of these determined processes in general and the nature of the initial state of the Roman system in particular. Most successful states are corrupted so soon by increasing wealth that they tend to collapse under its adverse consequences before they have a chance to develop into a universal empire; i.e. 'imploding' feedback loops swiftly start to accompany and then overshadow 'exploding' feedback loops. In the case of Rome, this process took more time because of the sticky character of its uniquely warlike mentality. This mentality remained intact during several centuries of increasing luxury. And this strong mentality, in its turn, had developed as a result of Rome's protracted struggles with its belligerent neighbours: 'If they had rapidly conquered all the neighbouring cities, they would have been in decline at the arrival of Pyrrhus, the Gauls, and Hannibal. And following the fate of nearly all the states in the world, they would have passed too quickly from poverty to riches, and from riches to corruption'.[53] So, the contingent initial state of the Roman system determined the long duration of the processes that governed its rise to world power and its subsequent fall.

I call Montesquieu's grasp of this process 'rigid' because his schematic distinction between *contingent initial state* and *subsequent determined processes* leaves little room for a convincing explanation of contingent events *in the course* of determined processes. For instance, why did the Byzantine Empire fall so extraordinarily late rather than much earlier?[54] This question caused

50 Montesquieu, *Considerations*, 169/Montesquieu, *Considérations*, xviii, 235.

51 Montesquieu, *Considérations*, xiii, 189–90.

52 Montesquieu, *Considérations*, xviii, 233.

53 Montesquieu, *Considerations*, 28–9/Montesquieu, *Considérations*, i, 96; cf. Montesquieu, *Considérations*, i, 90: 'Le Regne de Numa long & pacifique étoit très-propre à laisser Rome dans sa médiocrité; & si elle eût dans ce tems-là une territoire moins borné, & une puissance plus grande, il y a apparence que sa fortune eût été fixée pour jamais'.

54 Montesquieu, *Considérations*, xxiii, 278.

the author a lot of problems. The longevity of the Byzantine Empire implies an anomaly in the context of Montesquieu's determinist process explanations. Rome had managed to exist in spite of an ever-increasing corruption, because of its strong military core. But in the case of Constantinople the core was rotten from the start—hence, a priori, the prospects for this empire looked bleak. Why was the core of the Byzantine Empire rotten? This empire was the only country in the East with Christianity as the dominant religion. The cowardice, laziness, and feebleness of Asia was combined with a Christian bigotry that affected the military spirit of the empire. As an example, Montesquieu mentions Emperor Maurice (539– 602) who wept about the many lives that would be lost before he went to battle.[55] Moreover, Montesquieu notes that the feebleness of the Asians had made them more vulnerable to the influence of women. In most Asian nations this tendency had been counterbalanced very effectively by the institution of polygamy, which made it more difficult for any single woman to gain an ascendency on her husband. No such safeguard existed in the Byzantine Empire, where Asian weakness was combined with Christian monogamy, resulting for instance in the enormous power that accrued to Empress Theodora (500–548), a former prostitute who became the wife of Emperor Justinian (484–565). According to Montesquieu, Theodora constantly mixed the passions and the fantasies of her sex in affairs of state, and she consequently corrupted the victories of the empire.[56]

Given these inauspicious contingent beginnings, we would expect the subsequent phase, governed by determinist feedback loops, to be short. So how did the Byzantine Empire manage to hold on for such a long time? Montesquieu provides three explanations. Firstly, the possession of Greek fire allowed the Byzantines to set fire to invading fleets, especially Arab fleets coming from Africa and Syria. The Byzantines jealously guarded the secret of the chemical formula of this weapon and they successfully used it as its exclusive owners for many centuries. Secondly, thanks to its specific geographic position and its strong walls, Constantinople was left standing as the only wealthy city at a time when the Goths in the west and the Arabs in the east had destroyed most commerce and industry elsewhere. The Byzantines took over the silk industry from the Persians and their strong fleet protected their commercial interests. Thirdly, the barbarians who lived along the banks of the Danube happened to be significantly less ferocious than the barbarians who had caused the fall of the Western half of the Empire; these more pacific barbarians proved a useful

55 Montesquieu, *Considérations*, xxiii, 267–8.
56 Montesquieu, *Considérations*, xx, 253–4.

barrier against more aggressive barbarians.[57] Whether these are convincing answers is not relevant. What matters is that they have a *contingent* character. They are used to explain the Byzantine exception to a general rule: *inauspicious* contingent beginnings will be followed by a *short* determined process. Given Montesquieu's commitment to this rule, it made perfect sense for him to devote extra explanatory energy to the atypical case of the long-lived Byzantine Empire.[58]

4 Military and Political Context

So far it has been ascertained that: the *Considerations* provide a methodological exercise in causal explanation on the meso-level applied to the subject of the military rise and fall of Rome; Montesquieu describes Rome as a system with a contingent initial state that has a strong path-determining effect; and his use of self-reinforcing feedback loops explains how contingent and plastic initial configurations become highly determined and sticky in their subsequent operation.

My discussion so far has been of a formal and conceptual nature, but is not without relevance to an understanding of the unmistakably political agenda of the *Considerations*. It has already been noted that Montesquieu's discussion of Rome abounds with comparisons with historical events and situations in other ages. His favourite reference is to his own age and that of the reign of Louis XIV, which had ended only eighteen years before the publication of the *Considerations*. For instance, Montesquieu explains how the Romans managed to conquer and subsequently pacify Macedonia. The country was divided into four parts, contacts between these parts were discouraged; and the Macedonian aristocracy was transferred to Italy. Referring to Louis XIV he then remarks that 'if a great prince who reigned in our day had followed these maxims' he would have been much more successful.[59] Another critical reference to Louis XIV, more in particular to his introduction of tax rises in 1695 and 1710, is implied by a feedback loop that Montesquieu uses for the late Roman Empire: its citizens' ability to pay taxes decreased, which forced the state to

57 Montesquieu, *Considérations*, xxiii, 278–80.

58 See also Volpilhac-Auger, 'Nox', 397–8.

59 Montesquieu, *Considerations*, 71–2/Montesquieu, *Considérations*, vi, 137.

increase taxes, which decreased its citizens' ability to pay taxes even more, which weakened the Empire even more.[60]

Montesquieu's analysis of Rome's rise and fall allows him to drive home lessons about the strategic and political problems of his own country in his own age, and thereby to circumvent the censors of the Ancien Régime.[61] The military reverses of France against Britain during the War of the Spanish Succession (1701–1714) resulted in a bout of national soul searching by Montesquieu, Voltaire and other intellectuals.[62] Montesquieu had originally intended to publish the *Considerations* together with a much shorter essay, *Réflexions sur la monarchie universelle* ('Reflections on universal monarchy'), which contains a stinging condemnation of the expansionist wars of Louis XIV. He had a copy printed for his own use that combined the 'Universal Monarchy' with the *Considerations*, but in the end he decided against publication of the 'Universal Monarchy'. On the printed but unpublished copy he explained that he 'suppressed' the work, 'for fear that certain passages would be interpreted ill'.[63]

In the *Considerations* Montesquieu shows that warfare had still been a lucrative affair for the Romans, while the 'Universal Monarchy' makes the point that this was no longer the case in eighteenth-century Europe. Technical innovations such as firearms had matched the strength of various nations. Christianity had made the looting of enemy cities less acceptable. The invention of a postal system, the compass, and the art of printing had facilitated communication, and nations had become commercially entwined.[64] These factors all worked against military attempts at universal monarchy, as the failure of Louis had clearly demonstrated. The future belonged to countries whose strength rested on commercial wealth—in that sense Montesquieu agreed

60 Montesquieu, *Considérations*, xviii, 238 n. 17; see also Montesquieu, *Considérations*, ii, 102; vi, 137; xv, 203; xvi, 213; xxi, 263.

61 Cf. Oake, 'Montesquieu', 57; Andrivet, introduction in Montesquieu, *Considérations*, 19.

62 Rahe, *Montesquieu*, 1–60; see also Myers, 'Montesquieu', 47.

63 See Larrère, introduction in Montesquieu, *Réflexions*, ed. Weil, 336; see also Rahe, *Montesquieu*, 20 (English translation given here is by Rahe). It may have been Montesquieu's original intention to publish a monograph that would have comprised the *Considérations*, the *Réflexions*, and a third fragment on which he worked at the same time, on the mixed monarchy of England (see Carrithers, 'Montesquieu', 79; Lottes, 'Montesquieu', 115). The latter text would actually be published much later as the famous chapter xi.6 of the *Esprit*, 'On the English constitution' (see Rahe, *Montesquieu*, 40–2). Whether Montesquieu had indeed intended to publish this third text together with the *Considérations* and the *Réflexions* remains subject to debate (see Volpilhac-Auger, 'Montesquieu et l'histoire', 138 n. 11).

64 Montesquieu, *Considérations*, i, 95.

with De la Court and disagreed with Fénelon.[65] In the eighteenth century the model of Carthage stood a better chance of success than that of Rome. This formed the background for Montesquieu's admiring analysis of the English constitution. The new Carthage was superior to the new Rome.

The explanatory framework of feedback loops deployed in the *Considerations* is again vigorously at work in the 'Universal Monarchy', with parameters that provide Montesquieu with an explanation for the differences between Rome and eighteenth-century Europe. We have seen how in the *Considerations* Rome's successes produce wealth in the form of loot which in its turn feeds further successful expeditions. In the 'Universal Monarchy' the different commercial and financial parameters of eighteenth-century Europe produce a different picture. Commercial wealth allows nations to pay for professional armies and navies that are not dependent on patriotic citizens and not liable to a corruption of their mentality, if only because their soldiers and seamen already belong to 'the most despicable part of all the nations'.[66] Great wealth means greater military *potential*. But as soon as this potential is *actually* used for warfare, a corrective feedback loop will swiftly affect this military potential. When a monarch sends out his army to occupy a foreign province, he sends part of his treasure as well. Foreign provinces are no longer being looted as they were in classical times. The occupier pays for their subsistence, thus enriching the country that he has set out to conquer, and hence subsidising his own ultimate defeat.[67] Whereas it took centuries before Rome's military principles and mentality was 'corrected' by the feedback provided by wealth-generated moral corruption, the eighteenth-century European system provided a different correction to military adventures, one that was almost instantaneous. So here Montesquieu provides an explanation for the impossibility of a modern universal monarchy that, typically, involves both the general concept of a feedback process as well as a keen appreciation of changed systemic parameters.

5 Philosophical and Historiographical Context

In order to appreciate the particular and interesting character of Montesquieu's process explanations and feedback loops we need to consider the differences

65 See above, §1.2 and §2.5.

66 Montesquieu, *Réflexions* ii, 342; see also Montesquieu, *Considérations*, ii, 102; Montesquieu, *Esprit*, xi.6, 406.

67 See also Montesquieu, *Esprit* x.3, 378: 'La conquête est une acquisition; l'esprit d'acquisition porte avec lui l'esprit de conservation et d'usage, et non pas celui de destruction'.

between traditional Aristotelian and early modern mechanical (natural) philosophers in the generations before Montesquieu. When Aristotelian philosophers had tried to understand the rich world of colours and sounds of the natural world, they used no less than four different kinds of causes: material, formal, efficient, and final. They also had a wide and rich concept of motion, which was used to explain different kinds of change in general. The new mechanical philosophers on the other hand were reductionist. They tried to understand the physical world as perceived on the macro-level with a limited number of quantitative attributes of invisible corpuscles on the micro-level. Similarly, they narrowed the concept of motion to local motion only, expressed in terms of efficient causes only, while they dropped the other three Aristotelian causes. Many mechanical philosophers believed nature should be understood in secular terms without intrinsic aims provided by a divine creator—hence their polemics against final causes. The exclusive use of efficient causes went hand-in-hand with a determinist worldview: if a billiard ball is impacted by another billiard, then the trajectory and velocity of the impacting ball determines the trajectory and velocity of the impacted ball.[68]

All these concepts agreed well with the Scientific Revolution of the seventeenth century. In other areas they caused problems. For instance, mechanical determinism posed a problem for the belief in a human free will. René Descartes had tried to solve this problem with a strict separation between mind and body. Different things apply to minds and to bodies. Bodies are subject to the laws of physics, but minds are not; they are free. In the case of beings that have both a mind and a body, i.e. men, their body will be governed by the decisions of their mind. Sadly enough, Descartes did not explain how exactly this interaction between mind and body was possible. He left the answer to this difficult question to the next generation of mechanical philosophers. While the empiricist John Locke took great pains to defend a mechanical and determined world that was inhabited by free men, rationalist philosophers such Spinoza and Malebranche tended to defend efficient causality at the cost of human freedom. For these philosophers a belief in efficient causality amounted to a determinist worldview.[69]

In the midst of the radical changes implied by the new mechanical philosophy, the concept of substance proved remarkably resilient. Seventeenth-century mechanical philosophers redefined the qualities of substances, but substances themselves continued to hold sway. There was deep division about

68 Cf. Nadler, 'Doctrines of explanation', 513–52; on the term 'mechanicism' cf. Cohen, *Modern Science*, xxxvi.

69 See also Sleigh, 'Determinism', 1195–278.

the possibilities of knowledge of the essences of substances, and the empiricism of John Locke was as pessimistic as the rationalism of René Descartes had been optimistic; but all accepted the concept of stable substances with unchanging essences.

So here we have a mechanical background that can be summed up (admittedly somewhat simplistically) in four basic concepts: (1) static substances with unchanging essences; (2) reductionism; (3) determinism; (4) efficient versus final causality. Although Montesquieu was deeply influenced by each of these four concepts, his Roman history amounts to important novel conceptual contributions that are all related to his use of causal feedback loops on the meso-level of history.

Firstly, in Montesquieu's study of the extremely complex and dynamic processes of history, *immutable essences* are of very limited relevance. Yet commentators have continued to perceive essences in Montesquieu's work. Georges Benrekassa notes 'the persistent recurrence of a kind of global and overdetermining causality' that dominates the *Considerations* and tends towards the 'deployment of an essence'.[70] He adds that this tendency has the form of a causal idea that pertains to origins and that is especially pronounced in the first chapters of the work.[71] Catherine Volpilhac-Auger also notes that Montesquieu aims to 'extract from history what is really essential'.[72] I suggest a more precise, if still rather conditional use for 'essence' in the *Considerations*. The military principle of the Roman Republic that emerged as part of a feedback process in interaction with a competitive environment can be regarded as a sticky 'essence' that determined the historical process for ages to come. Montesquieu shared this belief in the possibility of the knowledge of essences with Aristotelian predecessors and mechanical philosophers of the Cartesian variety. But whereas these seventeenth-century philosophers had defined essences as attributes of material or spiritual *substances*, Montesquieu suggests an 'essence' that is the main property of a political and military *system*. This 'essence' had none of the general and unchanging nature of scholastic or mechanical essences of substances. The military 'essence' of the Roman Republic in the *Considerations* emerges in a feedback process that starts with a contingent initial state, and its stickiness is not of an absolute but merely of

70 Benrekassa, *La politique*, 59; see also Andrivet, introduction in Montesquieu, *Considérations*, 29–30.

71 Benrekassa, *La politique*, 60: 'pensée causale et pensée de l'origine'.

72 Volpilhac-Auger, preface in Montesquieu, *Considérations*, 33; she adds, in her preface in Montesquieu, *Considérations*, 33: 'Il ne s'agit pas pour autant d'établir pour lui-même un déterminisme historique, et surtout pas de chercher à prévoir les événements'.

a gradual nature. In the long term it evaporates, together with the system that it defines.

Secondly, Montesquieu's use of dynamic feedback processes not only amounted to a departure from the adherence to unchanging essences, but also implied a less rigid adherence to the *reductionist* ideal of his mechanical predecessors. For Montesquieu a commitment to efficient causality did not imply a commitment to reductionism. Sometimes he breaks the Roman system into its component parts, but just as often he does exactly the opposite, i.e. he is concerned to show the interaction between one part of the system with the other parts of the system, and also the interaction between the Roman system itself and its wider military and political environment. So whereas in the case of the mechanical philosophy efficient causes are part of a reductionist programme, Montesquieu's commitment to efficient causality is often expressed in the holistic sense that is typical for process explanations.

Thirdly, we have seen how a commitment to efficient causes seemed to dispose many mechanical philosophers in favour of *determinist explanations*, which are indeed strongly present in Montesquieu's *Considerations* as well. But can it be said that Montesquieu's determinist account of the process of *Roman* history was the expression of a *general* historical determinism? Not necessarily. Before the publication of the *Considerations* in 1733/1734 he had written two short works in which he considered the nature of causal explanation in history: *De la politique* (1725) and the *Réflexions sur la caractère de quelque princes* (1731–1733, 'Reflections on the character of some princes').[73] The first essay does not yet evince a firm commitment to either necessity or accident as primary forces of history, while the second essay, composed at the same time as the *Considerations*, actually marks a shift away from historical determinism. So there is no sign of a clear commitment to a determinist view *in general* prior to the publication of the *Considerations*. David Carrithers indeed makes the interesting suggestion that the determinism of the *Considerations* may have been prompted by Montesquieu's perception of inherent determinist characteristics in Roman history itself.[74] This suggestion of a 'local' determinism is

73 Montesquieu, *De la politique*, i, 112–19; Montesquieu, *Réflexions sur la caractère de quelque princes*, i, 519–531; see also Carrithers, 'Montesquieu', 77.

74 Carrithers, 'Montesquieu', 77: 'If one were ever tempted towards a determinist reading of history, Rome would be the example triggering that response. And Roman history, furthermore, was sufficiently distant in time from the eighteenth century to enable Montesquieu to discern a logical development in events that certainly would have escaped such a perception among those living through them or commenting on them in close proximity'. Carrithers, 'Montesquieu', 77: 'Certainly Montesquieu's Roman history cannot be characterized as embodying an actual transformation in his historical thought. Rather, Roman

compatible with Montesquieu's lack of interest in a general philosophy of history. His interest in the meso-level rather than the macro-level implied a preference for the analysis and comparison of individual political systems. He was interested as much in their differences as in their resemblances. The Roman system was highly determined. Other historical systems were less determined. In the *Considerations* Montesquieu deploys a conceptual apparatus that can be applied to various historical meso-systems *in general*. At the same time *differences* between these systems, including differences in the degree to which they are determined, are explained by variations in their initial states and variations in the nature and duration of their specific feedback loops.

In the *Spirit of the Laws* Montesquieu continued to develop the concept of local determinism. Work on this book started around 1739, two years after the publication of the *Considerations*. Montesquieu's mature masterpiece abounds with discussions about the interaction between different environments and different political systems, i.e. it abounds with topics on the meso-level that consequently have a potential for the use of local determinism. For instance, take the famous first chapter of the first book of the *Spirit of the Laws*, in which the author states that laws are *necessary* and follow from the nature of things. In this chapter on laws in general, Montesquieu assigns paradigmatic value to Newtonian physics and the law of gravity.[75] In this physical system, necessary and general laws follow from the nature of material bodies. At the same time Montesquieu observes that the intelligent world of human beings is not as well-governed as the material world. Human beings have a limited nature which makes them liable to error. It is in their nature to act 'by themselves'. Human beings do not follow the laws given to them by God or even the laws they gave to themselves. Human behaviour is partly determined, and partly non-determined. Moreover, in the same chapter Montesquieu provides us with an initial non-determined (contingent) stage followed by a subsequent determined stage. Before intelligent (human) beings existed, they were possible. Hence there existed *possible* relations between them and hence also *possible* laws. In his later *Défense de l'Esprit des lois* ('Defence of the Spirit of the Laws') Montesquieu repeats this point more specifically about the laws given to man by God. Before God created man, man had possible laws; only once man was actually created by God did these laws become necessary laws.[76] So, in the *Spirit of the Laws* and in its 'Defence' we see how Montesquieu continued

---

materials simply brought to the fore determinist sentiments traceable to *De la politique*'. But cf. Volpilhac-Auger, preface, to Montesquieu, *Considérations*, 34 n. 1.

75 Montesquieu, *Esprit* I.i, 233; see also Charrak, 'Le sens de la nécessité', 11.

76 Montesquieu, *Défense de l'Esprit des lois*, 1123; see also Charrak, 'Montesquieu', 15.

to use the conceptual combination of an initial contingent stage, followed by a subsequent determined stage. And he had first developed this combination when he tried to understand the military rise and fall of Rome.

Fourthly, like his mechanical predecessors, Montesquieu had no use for *final causes*. His causal feedback loops consisted exclusively of efficient causes. The Roman system had no inherent aim, nor was it itself part of any wider aim. Even when Montesquieu says that 'Rome was made for expansion, and its laws were admirable for this purpose', there is no need to read 'made for' and 'purpose' in a finalistic sense.[77] Montesquieu indeed ascribes to political leaders the ability to create the institutions of societies in their early stages, even though its leaders are subsequently formed by society. But in his account of the fiercely competitive environment of Latium, in which the political system with the best adapted feedback mechanism necessarily emerged on top, the words 'Rome was made for' form a formula that implies as little teleology as the conveniently succinct phrase 'The wings of an eagle are made for flying' implies teleology in the mouth of a modern evolutionary biologist.[78]

In order to appreciate the originality of Montesquieu's use of efficient (or non-final) causes for his analysis of Rome's military rise and fall, it is instructive to shift the background from mechanical scientists to historians who wrote before, during and after Montesquieu produced the *Considerations*. Historiography before Montesquieu was influenced by a tradition going back to Saint Augustine's *City of God* that dealt with history in providential terms and used final causes to give meaning to historical events. The main French representative of this tradition in the seventeenth century was Bishop Jacques-Bénigne Bossuet, whom we have already encountered as formidable enemy of Fénelon's quietism in the previous chapter.[79] Bossuet was court preacher to Louis XIV, a strong advocate of political absolutism and writer of the *Discours sur l'histoire universelle* (1681). Human history is constantly determined by the will of God. He directs human beings by their passions. Sometimes God bridles these passions and sometimes he gives them free room. If he wants to favour conquerors, he takes care that they are led by the passions of fear and horror. If he wants to create great legislators, he will curb their passions and give them the wisdom and the foresight that will prevent disasters striking their societies.[80] Bossuet does not deny that it is possible to discern clear patterns in human history. For instance, he uses the customs, *humeur*, and *tempérament* of

77 Montesquieu, *Considerations*, 94/Montesquieu, *Considérations*, ix, 158.
78 See also Richet, *Political Theory*, 56; Mason, 'Optimism', 204; Goyard-Fabre, *Montesquieu*, 38.
79 See above, §2.7.
80 Bossuet, *Discours*, 1025.

the Romans, for the explanation of such phenomena as the all-pervading jealousy between the Senate and people of Rome.[81] But history is first of all driven by the infallible decisions of God, in accordance with the rules of his own justice. If to us history may sometimes seem a random jumble, this is because of our ignorance of God's ends. What seems contingency to us, is actually a coherent design that contains all the causes and all the effects in the same divine order.[82]

While Montesquieu took care to maintain what has been described as 'a memorable silence' towards Christianity itself, he showed less restraint on Christian providentialism, and Bossuet is a favourite object of his derision in the *Considerations*.[83] He probably had Bossuet in mind when he made the following ironic remark about divine providence and the early successes of Islam against the Christians: 'God did not permit His religion to lose its predominance in so many places because He abandoned it, but because—whether its condition is one of outward humiliation or glory—it can always produce its natural effect, which is sanctification'.[84] Laws and principles of necessity should not be sought in heaven, but rather in 'the things themselves'. Conquest followed from Rome's own nature, not from any divine plan.[85] By eliminating final causes as an explanation for the military rise and fall of Rome, and by limiting himself to secular explanations in the form of efficient causes, Montesquieu paved the way for David Hume, William Robertson, Edward Gibbon, Voltaire, and other great eighteenth-century historians.[86]

Once Montesquieu had driven out explanations by divine providence in favour of feedback loops, many historians and philosophers of history could not resist the temptation to reintroduce final causes in the guise of the idea of Enlightened Progress. Voltaire's *Essay sur les moeurs* (1756) provides a general history of the gradual progress of human civilization and enlightenment. History had an emancipatory aim, and Voltaire used history as an enlightened

81 Bossuet, *Discours*, 990, 1003, 1013.

82 Bossuet, *Discours*, 949; see also Bossuet, *Discours*, 1024; cf. Bossuet, *Discours*, 953; Champailler, introduction in Bossuet, *Oeuvres*, 663; Ferreyrolles, 'Histoire', 191–2; Benrekassa, *La politique*, 47.

83 Shackleton, *Montesquieu*, 160; Perinetti, 'Philosophical reflection', 1107, 1117; Goyard-Fabre, *Montesquieu*, 32.

84 Montesquieu, *Considerations*, 201/Montesquieu, *Considérations*, xxii, 265; see also Montesquieu, *Considérations*, xxiii, 280: 'C'est leur felicité que Dieu ait permis qu'il ait dans le monde des Turcs & des Espagnols, les hommes du monde les plus propres à posseder un grand Empire'.

85 Goyard-Fabre, 'Rome', 38.

86 Mason, 'Optimism', 205.

weapon in the struggle against ignorant superstition and the furtherance of enlightenment.[87] Similar contentions can be found in Condorcet and in Kant. The latter posited the regulative idea of providence or providential nature that showed the way towards the goal of the human race. In his *Mutmaßlicher Anfang der Menschengeschichte* (1796, *Conjectures on the Beginning of Human History*) Kant wrote that history is the story of man's transition from a rude and purely animal existence to a state of humanity, from the call of instinct to the guidance of reason, and from the guardianship of nature to a state of freedom. In that sense, history is a story of progress towards human perfection.[88] While Kant and other enlightened historians thought in terms of linear progress, Montesquieu's idea of history was rather cyclic—hence his affinity with Vico.[89] Montesquieu did not believe in endless progress for humanity as a whole, but in the likelihood of birth, rise, and decline of separate civilizations. Here again he did not posit one single paradigm on the macro-level of history, but rather single processes on the meso-level, with Rome as his favourite example. Montesquieu did not only ridicule the providential history of Bossuet, but also the modern enlightened prophets of progress. He thought that Voltaire would never be a good historian because he was like a monk who did not write for the sake of historical topics themselves, but for the glory of his enlightened ideology: 'Voltaire writes for his convent'.[90]

So, Montesquieu's explanations for the military rise and fall of Rome in the *Considerations* amounted to more than a military history. His dynamic interpretation of the traditional concept of substances with unchanging essences; his holistic use of explanatory processes; his model of contingent initial conditions followed by determinist feedback loops; and his consistent use of efficient causes in the teleological limbo that superseded providential finality and was then followed by enlightened finality, allowed him to make a vital contribution to modern historiography.[91]

---

87 Mason, 'Optimism', 206–10; Senarclens, Montesquieu, 126.
88 Kant, *Conjectures*, 221–3; see also Carrithers, 'Montesquieu', 79–80.
89 See above, §3.2.
90 Quoted in Senarclens, 'Montesquieu', 127.
91 See also Lottes, 'Montesquieu', 115.

CHAPTER 4

Carl von Clausewitz on Limited War: a Three-Stage Interpretation

1 Introduction

Carl Philip Gottfried von Clausewitz was born in 1780 in Burg in the Kingdom of Prussia. He entered the Prussian army in 1792 and saw his first action in 1793 in the turmoil that swept over Europe after the outbreak of the French Revolution. After France and Prussia had signed a separate peace, Clausewitz was regimented near Berlin. In 1801 he was admitted to the Institute for Young Officers, where he attended lectures in philosophy, history, and military science and theory. He became *aide-de-camp* to Prince August, nephew of King Frederick William III, whom he accompanied in the battle of Auerstäd. Near this place and near Jena the French utterly crushed the Prussian army in 1806. Clausewitz became a captive of the French and after his release was passionately involved with the reform movement that tried to regenerate the forces of his defeated nation. In 1812, the year of Napoleon's ill-fated invasion of Russia, he transferred his services to the Russian army. He participated in the Wars of Liberation in 1813–1814 and also in the Waterloo campaign of 1815. He probably started work on *On War* around 1816. The Prussian King never quite forgave him the temporary transfer of his allegiance to the Tsar, but in 1818 he was appointed head of the Prussian War College and promoted to the rank of brigadier general. In this period he also produced numerous historical writings on the campaigns of Napoleon. In 1830 his intellectual pursuits were interrupted by his appointment as chief of staff to the army that was formed for the eventuality of an intervention in Poland. He fell victim to the great cholera outbreak in 1831.[1]

Clausewitz's distinction between absolute and limited war is vital for his strategic thinking. His distinction still figures prominently in modern scholarship.[2] One topic in contemporary debate is that of the differences and similarities between Clausewitz's concept of absolute war and the twentieth-century concept of total war. In the present chapter I will discuss another question: how

---

1 Schuurman, 'Clausewitz on real war', 372.

2 See Aron, *Penser la guerre*, I, 264–6; Gat, *Military Thought*, 224–5; Cooker, *Rebooting Clausewitz*, 42–3.

 DOI:10.1163/9789004536678_006

exactly did Clausewitz's views on limited war develop? I propose a new model and I argue that his views in *Vom Kriege* (*On War*)[3] on the constraints on warfare evolved in three continuous *stages* that can be defined by two system theoretical *dimensions*. The two dimensions are the absence or presence of *interaction* (*Wechselwirkung*) between human actors in the form of negative and positive causal feedback loops; and the absence or presence of a *holistic perspective* in the form of a political system that forms the context for the military subsystem. The first stage guided by these two dimensions is *friction* as a constraint on the effectiveness of the execution of military plans on paper. In this initial stage both the interactive and the holistic perspective are switched off. The second stage consists of *suspension* as a constraint on the intensity of fighting in battles and campaigns. In this stage the holistic perspective remains switched off, but the interactive perspective is switched on. In the third stage *political objectives* act as a constraint on military objectives. In this final stage both the interactive and the holistic perspective are switched on.[4]

Clausewitz's *On War* was an unfinished work in progress at his death in 1831. Any attempt at a reconstruction of the various stages of his military thinking should take into account the order in which *On War* was written. In 2014 Paul Donker discovered two previously unknown drafts of chapter 1 to 3 of book I and the complete manuscript of books I and II in the *Wehrtechnische Studiensammlung* in Koblenz, Germany. He is currently using these manuscripts for a detailed reconstruction of the genesis of *On War*. At present his assumption is that Clausewitz started the revision of books VI, VII, and VIII already in 1826 (not in 1827); that he revised the first three chapters of book I after 1827; that the famous chapter I of this book was written last; that parts of book VIII were revised shortly before or even simultaneously with the revisions of book I; and that the revision of book II was separate from the revision of the other books.[5] Besides *On War* I will use various less well-known essays by Clausewitz to shed light on the development of his thinking.

3 Carl von Clausewitz, *Vom Kriege*, ed. Werner Hahlweg (18th ed., Bonn: F. Dümmler, 1973); translation: Carl von Clausewitz, *On War*, ed. and transl. Michael Howard and Peter Paret (Princeton, New Jersey: Princeton University Press, 1984).

4 For the system theoretic dimensions see also Gat, *Military Thought*, 221–3; Beyerchen, 'Clausewitz', 67; but cf. Roth, 'Chaos', 30–44; Holmes, 'Planning versus Chaos', 130. Finally, Beckmann, *Clausewitz* uses systems theory to analyse Clausewitz, but is not primarily interested in the development of his thinking on constraints on warfare.

5 Email by by Paul Donker from 13 November 2020; see also Donker, 'The Genesis', 101–17; Donker, 'The Evolution'. Cf. Schering, *Wehrphilosophie*, 250–255; Aron, *Penser la guerre*, 1, 96–107, 177–8, 217 n. 1, 238; Gat, *Military Thought*, 257–65; Diniz, 'A Criterion', 17–22.

2 Interaction and Holism

Clausewitz's early interest in the concept of interaction is not surprising, since this concept is at the very heart of warfare itself. In the early *Leitfaden zur Bearbeitung der Taktik der Gefechtslehere* ('Outline to work on the tactics of the doctrine of combat') he first discusses everything that can be learned about military forces separately and he then continues with a substantial separate section on the interaction (*Wechselwirkung*) between the plans and actions of two enemies.[6] A concern with the relation between whole and part, i.e. a holistic perspective, is present as well. These interests are continued in *On War* itself. In book II, on the theory of war, the author notes the unpredictable character of the contact between opponents and he stresses that this is due to 'the very nature of interaction'.[7] Chapter 7 of the long book VI on defence is completely devoted to the interaction between attack and defence.[8] In the *Strategie aus dem Jahre 1808* ('Strategy from the year 1808') Clausewitz explains the meaning of tactics in the context of the meaning of strategy. Tactics is the use of military forces in battle, and strategy is the use of separate battles for the aims of war. Politics provides the aims for strategy which in its turn provides the aims for tactics.[9] Moreover, often the relation between whole and part itself is expressed in terms of interaction, i.e. often the two basic dimensions are combined. In book III the author stresses the interaction between the various parts of strategy, which can only be distinguished in thought while forming one single whole in reality.[10] In book IV, after chapters devoted to the engagement in its 'absolute' i.e. in its separate sense, he turns to its relations with the larger whole, and it is only this perspective that allows him to analyse the meaning and significance of individual engagements.[11] At the start of book V, he explains that numerical strength, organization, maintenance, and

6 Clausewitz, *Leitfaden*, 1169: 'die Verwendung der Streitkräfte im Gefecht aus ihrer Natur selbst'. The 'Outline' was included by Marie von Clausewitz in vol. 3 of *Vom Kriege* (Berlin: F. Dümmler, 1834), 281–6 and by Hahlweg in Clausewitz, *Vom Kriege*, 1103–80. The 'Outline' may be one of the products of Clausewitz's 'Forschungen und Bestrebungen' following his appointment as teacher at the General War College and tutor to the Crown Prince in 1810; see Marie von Clausewitz, preface in Clausewitz, *Vom Kriege*, 174.

7 Clausewitz, *On War*, 138/Clausewitz, *Vom Kriege*, ii.2, 288; see also Clausewitz, *On War*, 127/ Clausewitz, *Vom Kriege*, ii.1, 269; Clausewitz, *On War*, 136/Clausewitz, *Vom Kriege*, ii.2, 283.

8 Clausewitz, *On War*, 377/Clausewitz, *Vom Kriege*, vi.7, 645.

9 Clausewitz, *Strategie aus dem Jahre 1808*, 33, 58, 62–63.

10 Clausewitz, *On War*, 183/Clausewitz, *Vom Kriege*, iii.2, 355; see also Clausewitz, *On War*, 192/Clausewitz, *Vom Kriege*, iii.6, 370; Clausewitz, *On War*, 203/Clausewitz, *Vom Kriege*, iii.10, 387.

11 Clausewitz, *On War*, 236/Clausewitz, *Vom Kriege*, iv.5, 436.

other aspects of military forces, are all interconnected, although each must be examined as a separate entity for the sake of an orderly analysis.[12] Finally, at the start of chapter 27 of the same book, Clausewitz observes on the one hand that in war, 'more than anywhere else, it is the whole that governs all parts', while on the other hand, without a step-by-step discussion from the simple to the complex 'we should have been swamped by a multitude of vague concepts; more particularly, the variety of interactions that occur in war would have constantly confused our ideas'.[13] So, interaction and holism are already present in Clausewitz's earliest thinking about war. Yet both dimensions remain firmly switched off in his account of friction.

3 Friction

Clausewitz's earliest and most comprehensive discussion of the factors that act as a constraint on warfare is that of friction in an essay written in 1812, *Die wichtigsten Grundsätze des Kriegsführens* ('The main principles of warfare').[14] In her preface to *On War*, Clausewitz's wife Marie von Brühl explains that the 'Principles' were the result of his lessons to the Prussian crown prince Friedrich Wilhelm (1795–1861) and his new job at the General Military Academy of Berlin (1810). She notes that the essay contains the 'seeds' (*Körner*) of *On War*.[15] Under the header of 'strategy' Clausewitz formulates a number of general objectives that should typically be achieved through swift, aggressive, and concentrated military action. Both the objectives and their achievement are formulated in a number of principles.[16]

Clausewitz concludes his essay with the observation that these principles are all very easy, and that they can be grasped with very little academic proficiency. Yet it is surprisingly difficult for a commander to adhere to these principles. This is because warfare is similar to 'a complicated machine with a tremendous friction', so that plans on paper can be executed in reality with the

12 Clausewitz, *On War*, 279/Clausewitz, *Vom Kriege*, v.1, 499; see also Clausewitz, *On War*, 292/Clausewitz, *Vom Kriege*, v.5, 518.

13 Clausewitz, *On War*, 484/Clausewitz, *Vom Kriege*, vi.27, 807; see also Clausewitz, *On War*, 484–485/Clausewitz, *Vom Kriege*, vi.27, 808; Clausewitz, *On War*, 523/Clausewitz, *Vom Kriege*, vii.1, 869; Clausewitz, *On War*, 577/Clausewitz, *Vom Kriege*, viii.1, 949; Aron, *Penser la guerre* 1, 187; Gat, *Military Thought*, 237.

14 Clausewitz, *Die wichtigsten Grundsätze*, 1047–86; see also Paret, *Clausewitz and the State*, 202; Gat, *Military Thought*, 187 n. 52; Watts, *Clausewitzian Friction*, 5–8.

15 Marie von Clausewitz, preface in Clausewitz, *On War*, 66/Clausewitz, *Vom Kriege*, 174.

16 Clausewitz, *Grundsätze*, 1070–971.

greatest effort only.[17] Clausewitz then proceeds to give eight rather disparate reasons for friction that can be both of a psychological and of a physical nature. His list includes insufficient information, false rumours, demoralization, and logistical problems. All these factors have an adverse effect on the internal functioning of the complicated military machinery. These internal problems have various forms, but the main problem is that 'the free will, the spirit of the commander is hampered'.[18] Friction is very much a problem for the commander, and it should also be solved by the commander, 'who with bravery and strong determination follows his purposes with a strong ambition in spite of everything'.[19] Since the 'Principles' were written for the crown-prince, i.e. the future commander-in-chief of the Prussian army, this perspective is not at all surprising.

The perspective of Prussian military leadership may also throw light on the absence of the topic of friction from the *Strategie aus dem Jahre 1804* ('Strategy from the year 1804') and on its prominent place in the 'Principles' from 1812.[20] Friction is basically everything that can derail the military operations planned by commanders. Between 1804 and 1812 operations went very wrong indeed for the Prussian army, which was crushed by Napoleon during his military campaigns of 1806 and 1807. The first time that Clausewitz uses the word *Friktion* was precisely in this context, in a letter to Marie of 29 September 1806. In this letter he writes about the immense problems, caused by incompetence and internal division, that his teacher Gerhard von Scharnhorst (1755–1813) was trying to overcome at the Prussian military headquarters. He complains that Scharnhorst's talent is being wasted, 'when it is broken by the many obstacles formed by convention and when it is paralyzed by a continuous friction of different opinions'.[21] Clausewitz would gain first-hand experience of these

17 Clausewitz, *Grundsätze*, 1080.

18 Clausewitz, *Grundsätze*, 1080.

19 Clausewitz, *Grundsätze*, 1083.

20 Clausewitz, *Strategie aus dem Jahre 1804*, 3–45; see also Kessel, 'Zur Genesis', 408–9 and Gat, *Military Thought*, 187 n. 52.

21 Schwartz, *Leben*, 1, 224–5: 'Unter wie schwierigen Umständen dieser Mann wirkt ist kaum zu glauben; man erhält davon einige Vorstellungen, wenn man weiß, daß *drei* Feldherren und *zwei* Generalquartiermeister sich bei dieser Armee befinden, da doch nur *ein* Feldherr und *ein* Generalquartiermeister da sein sollten. Ich bin meinem Leben noch nie auf einen Menschen gestoßen, der mehr geeignet gewesen wäre, Schwierigheiten der Art zu besiegen, als der Mann, von dem ich hier rede; allein wie wiel muß nicht von der Wirkungen des Talents verloren gehen, wenn es sich an so vielen Hindernissen der Convenienz bricht, wenn es durch eine unaufhörliche Friction fremder Meinungen gelähmt wird. So viel ist gewiß, daß ein unglücklicher Ausgang, wenn er uns Treffen sollte, allein Folge dieser kleinlichen Convenienz-Rücksichten ist'.

problems himself. It would not be difficult to find examples for each of the eight points mentioned in the 'Principles' in the many disasters that befell the Prussian command during these fateful years.

Much of Clausewitz's discussion of friction in the 'Principles' reappears in a draft for *On War*, in the earliest chapters of book I of *On War* itself and, with decreasing frequency, in chapters that were written at a later stage.[22] In I.7 friction is again mentioned as cause of the difference between plans on paper and real war. At this early stage friction is even considered to be *the only* factor responsible for this disparity.[23] The solution of this problem is again sought in leadership. A great general will attain his objectives in spite of friction, even if it destroys his own army. For this he needs a force of character that is often compared to a rock or an obelisk: 'Iron will-power can overcome this friction; it pulverizes every obstacle, but of course it wears down the machine as well (...) The proud spirit's firm will dominates the art of war as an obelisk dominates the town square on which all roads converge'.[24]

The concept of friction forms Clausewitz's first explanation for constraints on warfare. It is obvious that the perspective of a political context (i.e. the first dimension of my model) is non-existent in this early account. The interactive context (i.e. the second dimension) is even explicitly *negated*. The whole point of great leadership, in an anti-frictional context, is precisely the *prevention* of any undesired interaction between the military machine and its surroundings. A battle-hardened army *acts* on the command of its general but does not *react* to the debilitating influences of its surroundings, nor does the general — hence Clausewitz's stone-like imagery for ideal leadership.

4 Suspension

In chapter III.16 of *On War*, 'The Suspension of Action in War', Clausewitz embarks on a new account of the factors that act as a constraint on warfare. If the nature or essence of war consists of mutual destruction, then we would expect the opponents to be constantly engaged in adverse action. This implies a contrary symmetry that Clausewitz calls 'polarity'. The implication of true polarity is that every advantage gained by one side would be a precisely equal

22 See Clausewitz, 'Friktion im Kriege' in *Niederschriften des Werkes "Vom Kriege"*, in Clausewitz, *Schriften*, ii.1, 639–41.

23 Clausewitz, *On War*, 119/Clausewitz, *Vom Kriege*, i.7, 262; cf. Aron, *Penser la guerre* 1, 200.

24 Clausewitz, *On War*, 119/Clausewitz, *Vom Kriege*, i.7, 261–2; see also Clausewitz, *On War*, 117/Clausewitz, *Vom Kriege*, i.6, 259; Rogers, 'Clausewitz', 1175.

disadvantage to the other. Hence polarity invites both opponents to increase their efforts. Warfare tends to extremes. The Napoleonic wars showed that war can indeed reach the high levels of 'energy' implied by polarity—but this is not always the case. Actually, history shows that this forms the exception to a rule that consists in 'the tame conduct' of conflicts 'in which the hostile spirit of true war is held in check'.[25] Most wars are interrupted by sustained periods of inactivity by both parties, i.e. pauses, which are not to be confused with the more conclusive termination of a conflict by peaceful or forceful means. These pauses 'function as inherent counterweights and prevent the clockwork from running down rapidly or without interruption'.[26]

Clausewitz then provides three causes for this phenomenon of suspension of action (*Stillstand*) in war. 'The first of these, which creates a permanent tendency toward delay and thus becomes a retarding influence, is the fear and indecision native to the human mind. It is a sort of moral force of gravity, which, however, works by repulsion rather than attraction: namely, aversion to danger and responsibility'.[27] This first cause is clearly a form of friction, based on a human emotion that Clausewitz had already discussed in the 'Principles': fear in the face of momentous decisions. The friction-like nature of this first explanation in III.16 is confirmed by the typical dead-weight characterization, 'force of gravity' (*eine Art Schwere*), and also by the remedy of a commander with 'an enterprising martial spirit'.[28]

Something new announces itself in the same chapter III.16 when the author discusses the second and third causes of suspension. Here he starts to include the effect of *interaction* between actors—primarily the interaction between commanders through the medium of their respective military machines. This second cause of suspension starts like a form of friction: inaction caused by ignorance and mistakes about the strength and the whereabouts of the opponent. But Clausewitz then continues with the observation that both opponents often suffer from the same problem, with the result that they will both feel too weak to take the initiative and start waiting for each other, which for all its obviously passive nature still amounts to a clear form of interaction.[29] Reliable and precise information about one's *own* forces is difficult to obtain, let alone information about the *opponent*. As a result, both opponents are caught in a web of assumptions and conjectures. Often it is only after the conclusion of a

25 Clausewitz, *On War*, 218/Clausewitz, *Vom Kriege*, iii.16, 409.
26 Clausewitz, *On War*, 217/Clausewitz, *Vom Kriege*, iii.16, 408.
27 Clausewitz, *On War*, 217/Clausewitz, *Vom Kriege*, iii.16, 408.
28 Clausewitz, *On War*, 217/Clausewitz, *Vom Kriege*, iii.16, 408.
29 Clausewitz, *On War*, 217/Clausewitz, *Vom Kriege*, iii.16, 408–9.

campaign that things become clear.[30] Hence mutual passivity during a campaign is a frequent phenomenon.

Interaction plays a similar role in the third cause of the suspension of military actions. This cause is a 'ratchet-wheel' that occasionally causes a complete halt in the machinery, and it consists in the greater strength of the defensive over the offensive. This means that if A does not feel strong enough to attack B, it does not follow automatically that B will be strong enough to attack A. The greater strength of the defensive creates an interactive vacuum in which the parties are waiting for each other. This vacuum frequently proves an insurmountable barrier. The opponent who switches from defence to attack *loses* the advantage of defence, but his opponent will *gain* this advantage. In this way, there will be a double effect. The result will often be that both opponents will not only *feel* too weak but actually *are* too weak for an offensive. So, both in the second and the third case the result is inaction.[31]

While it is thus possible to distinguish between the absence of interaction in the older concept of friction in the first explanation and the presence of interaction in the second and third explanation in *On War* III.16, this difference is even more pronounced in Clausewitz's essay *Ueber das Fortschreiten und den Stillstand der kriegerischen Begebenheiten* ('On the Progression and Suspension Of Military Affairs'), appended to a letter written to August von Gneisenau (1760–1831) on 4 March 1817.[32] The author himself describes 'Progression and Suspension' as a 'preliminary work' (to *On War*) and advises its recipient to throw it in the fire as 'left shavings', an advice that was obviously and fortunately neglected.[33] This paper contains exactly the same three points as chapter III.16 of *On War*, but the differences that set apart the first point from the two subsequent points in III. 16 are expressed more vigorously. Clausewitz compares a commander's 'natural fear of an all-decisive ruling of fate' with a card player who is loath to gamble all his money in one single stake, although there is no objective reason why multiple smaller gambles are preferable to a single large gamble.[34] The reason for this fearful behaviour lies 'in man himself', i.e. as in III.16, the first point lacks an interactive dimension. When Clausewitz continues with the two other points the perspective changes in precisely this interactive way. He no longer talks about 'one commander but rather of both together'.[35] He tries to describe 'the nature of war' and presents

30 Clausewitz, *On War*, 165/Clausewitz, *Vom Kriege*, ii.5, 326–7.

31 Clausewitz, *On War*, 217–18/Clausewitz, *Vom Kriege*, iii.16, 409.

32 Clausewitz, *Ueber das Fortschreiten*, ii.1, 248–55; see also Kessel, 'Zur Genesis', 413–14.

33 Clausewitz, *Ueber das Fortschreiten*, 243.

34 Clausewitz, *Ueber das Fortschreiten*, 248–9.

35 Clausewitz, *Ueber das Fortschreiten*, 250.

the two interactive factors that he would also give in III.16: limited knowledge about the opponent and superiority of the defensive. Interaction is especially hinted at in the second explanation: 'In this way, when a *true* motive of one commander meets an *imagined* motive of the other, a real suspension can happen, without any contradiction to the nature of the thing'.[36] The author's interest in 'the nature of the thing' points to another intriguing difference: while the first cause is accompanied with the standard remedy for overcoming friction in the form of energetic leadership, no such remedy is formulated in the second and third points, where the emphasis is squarely on analysis rather than prescription, which is indicative of things to come in *On War*.

If the first explanation of suspension in warfare is indeed based on an older concept of friction while the second and third explanations in III.16 are newer, then it is not surprising that we see only the second (limited knowledge) and the third (superiority of the defensive) reappear in the most recent part of *On War*, i.e. chapter I.1. Here superiority of the defensive is placed in I.1.17, while limited knowledge reappears in section I.1.18. So their order has been reversed, no doubt because Clausewitz has come to regard the former as the cause that explains 'most periods of inaction that occur in war'.[37] This evaluation may have been influenced by his work on the very long book VI, devoted to defence and written after III.16 but before I.1.17–18.

The special character of the second and the third explanations of suspension can be understood in terms of causal feedback loops. We have already seen that *positive feedback* occurs when both the input and the output of a system increase or when both the input and the output decrease.[38] We have also seen when that *negative feedback* occurs when part of an increasing output is used to decrease the input, or vice versa. And while uncorrected positive feedback leads to an explosion or implosion of the system, negative feedback has a balancing effect that will tend to preserve the system. The corrective action of this mechanism never occurs instantaneously and always takes time. A thermostat will only switch on the heating after the temperature falls below a certain value. Without this measurable decrease in temperature the thermostat would never be prompted to send its corrective signal; so the subsequent increase in temperature will demand some time. This delay causes the typically oscillating pattern of a negative feedback system.

It is this *concept*, if not the *terminology*, of a negative causal feedback loop that seems to inform Clausewitz's second cause of suspension as mentioned in

36 Clausewitz, *Ueber das Fortschreiten*, 250–1, Clausewitz's italics.

37 Clausewitz, *On War*, 84/Clausewitz, *Vom Kriege*, i.1.17, 205.

38 See above, §3.3.

chapter III.16, and later in section I.1.18: imperfect knowledge of the military situation that can lead a commander to believe that the initiative lies with the enemy while in fact it lies with him. In theory, misjudged initiatives could lead as easily to an increase in action as to a decrease. Clausewitz points out that in practice, however, since men are by nature prone to an overestimation of the forces of their enemy, imperfect intelligence about the enemy tends to favour inaction over action: 'The possibility of inaction has a further moderating effect on the progress of the war by diluting it, so to speak, in time by delaying danger, and by increasing the means of restoring a balance between the two sides'.[39] In due time, a commander who is in a position to attack may receive the information that will prompt him to take this action after all, with the attack being called off again when the information has changed.

A similar account can be given of Clausewitz's third cause of suspension: the advantage of the defensive over the offensive. In his more extensive discussion in section I.1.17, Clausewitz repeats that true polarity between two active parties is elusive and he stresses the importance of the defensive: 'that the impulse created by the polarity of interests may be exhausted in the difference between the strength of attack and defence, and may thus become inoperative'.[40] In this way, the advantage of defence over attack explains why wars can be interrupted by long periods of inactivity. A commander will only switch on the system when his offensive strength outweighs the defensive strength of his opponent, and the system is switched off as soon as this condition no longer exists.

So in both cases Clausewitz seems to assume a mechanism that switches on and off, depending on whether certain relevant values are outside or inside a specific bandwidth. This system functions irrespective of the precise nature of the values. Subjective strength perceptions and objective differences between the offensive and defensive have the same effect. The system will remain active only as long as certain threshold values outside a specific bandwidth are maintained. The negative feedback mechanism will cause a shut-down as soon as these values fall back within the specific bandwidth. This notion of a passive bandwidth is expressed eloquently in the essay 'Progression and Suspension': 'So there is a certain space (*ein gewisser Raum*) between defensive and offensive power that is situated like a neutral territory between the two armies and that often produces suspension, because it can be crossed by neither army'.[41] A will only attack B if the 'space' consisting either in a lack of information or a lack of offensive strength is 'crossed'; only then will the

39 Clausewitz, *On War*, 85/Clausewitz, *Vom Kriege*, i.1.18, 206.

40 Clausewitz, *On War*, 84/Clausewitz, *Vom Kriege*, i.1.17, 205.

41 Clausewitz, *Ueber das Fortschreiten*, 251.

mechanism switch on, and only then will offensive action terminate a phase of suspension. Here and elsewhere, Clausewitz's writing seems infused with mechanical analogies. Devices for the implementation of negative feedback control had been invented ages before.[42] More specifically, Clausewitz could draw on the very recent and well-known invention (or improvement) by James Watt (1735–1819) of the centrifugal speed governor for steam engines; this device provided a negative feedback loop with which to control the engine speed.[43]

We have seen that an important aspect of negative feedback loops is their delayed correction pattern. Clausewitz showed an acute awareness of this mechanism in a military context. The opponents in a campaign constantly need time to decide about the continuation of offensive or defensive action. In chapter I.1.19 Clausewitz writes that decisions about the continuation or cessation of military action are based on a calculation of probabilities, and 'the amount of time available for such calculation will depend on the pace with which operations are taking place'.[44] The dramatic consequences of this mundane fact transpire in his concept of culmination. In chapter VII.5 Clausewitz points out that some strategic offensive operations are directly terminated with a peace treaty, but that most offensives are continued until the attacker has reached a point where his strength would barely suffice to maintain a defensive position in which he waits for peace. He explains that most offensives are continued until the attacker has reached a point where his strength would just barely suffice to maintain a defensive position: 'Beyond that point the scale turns and the reaction follows with a force that is usually much stronger than that of the original attack. This is what we mean by the culmination point of the attack'.[45] The occurrence of a strong reaction after an offensive action has passed the culmination point is also expressed elsewhere, for example in chapter III.17, i.e. precisely the chapter that connects the introduction to kinds of suspension in the form of negative feedback in III.16 with an analysis of the dynamic character of this suspension in III.18. In this connecting chapter III.17 the author discusses the case where the offensive power of Napoleon's army had exhausted itself in the disastrous campaign against Russia in 1812. In the following year the defensive power of Russia and also Prussia switched 'with enormous energy' to the offensive.[46]

42 See above, §3.2 n. 38.

43 Wellstead, 'Engine Speed Control', 1–2; see also Mayer, 'The Origins', 117–18.

44 Clausewitz, *On War*, 85/Clausewitz, *Vom Kriege*, i.1.19, 207.

45 Clausewitz, *On War*, 528/Clausewitz, *Vom Kriege*, vii.5, 639; see also Kessel, 'Zur Genesis', 407–8.

46 Clausewitz, *On War*, 220/Clausewitz, *Vom Kriege*, iii.17, 412.

Why, according to Clausewitz, is it that an attacking army can be so utterly destroyed after it has gone beyond the point where it could still save itself by switching to a defensive position? Firstly, there is the point that the defence is the stronger form of warfare; and during a campaign the forces of the attacker will often decrease while those of the defender will increase, until a tipping point is reached. Secondly, once our train of thought is set in a certain direction, it tends to continue in that direction, especially if time and occasion for a careful reconsideration is lacking—for instance when we are in the midst of a battle. Preoccupied by intense and dramatic action, the attacker frequently fails to apprehend the culmination point. Clausewitz makes the comparison with a horse pulling a load uphill, which finds it easier to go on than to stop. So the explanation is a kind of psychological force of inertia. A good general will not make this mistake himself and he will be able to profit promptly from this mistake when it is committed by his opponent.[47]

5 Politics

So far, we have seen that the first phase of Clausewitz's thinking consists of friction, in which both the interactive and the holistic perspective remain switched off. We have also seen how in the second phase, consisting of suspension, the interactive perspective is switched on. This interactive perspective has the form of negative feedback loops. Finally, in the third phase, Clausewitz switches on the holistic perspective as well. This time interaction has the form of positive rather than negative feedback loops. The holistic perspective has the form of political objectives that determine the objectives of military action.

Quite apart from his military and conceptual *preferences*, Clausewitz had always appreciated the apparent *fact* that different historical eras show difference in the objectives, means, scope, and intensity of military action. In chapter III.17 of *On War*, 'The Character of Contemporary Warfare', he describes how 'Bonaparte's audacity and luck have cast the old accepted practices to the winds. Major powers were shattered with virtually a single blow'.[48] He also observes how in that epic age entire nations were mobilized, that Prussia's use of militias in addition to its standing army increased its strength by a factor six, and that wars on this scale obviously have to be fought along different principles than wars that are fought only by standing armies. The word

---

47 Clausewitz, *On War*, 572–3/Clausewitz, *Vom Kriege*, viii, 944–5; see also Vego, *Joint Operational Warfare*, 73–95.

48 Clausewitz, *On War*, 220/Clausewitz, *Vom Kriege*, iii.17, 412.

'contemporary' (*heutig*) in the title of the chapter implies that for Clausewitz warfare in the Napoleonic style still has a paradigmatic value. At the same time the word 'contemporary' hints at a possible awareness that Napoleonic warfare might prove a merely transient phase in military history. This point is indeed made more explicitly, if ironically, in a later chapter: 'It is quite possible that at some time in the future, Bonaparte's campaigns and battles will be considered brutalities, almost blunders, while the old-fashioned dress sword of antiquated and desiccated manners and institutions will be relied upon and praised'.[49] Clausewitz started his work on *On War* during the long interlude between the Napoleonic Wars and the Wars of German Unification half a century later. In this period the dramatic multiplier effects of industrialisation and technological developments had not yet made themselves felt on the battlefield. Hence for Clausewitz a reversion to eighteenth-century warfare was a distinct possibility.

Since Clausewitz understood that Napoleonic warfare demanded different 'principles', it is not surprising that, conversely, he thought that a return to more limited forms of conflict should be explained by a proper theory as well. In the earlier-mentioned essay 'Progression and Suspension' of 1817, he already perceives a need to provide this explanation for what might appear to be a new military trend—although so shortly after Waterloo he is still confused: 'The campaigns in which the rapid and continuous course of blood-stained action took place, in which the clockwork went off as fast as natural friction allowed (e.g. 1706, 1757, almost all the campaigns of *Bonaparte*), have become so rare and are accompanied by such special circumstances, that it is unclear whether they should be regarded as a rule or as an exception, i.e. as the opposite of a rule, which is a fine alternative (*eine tolle Alternative*) for a theory'.[50] So, here we have a clear and early realization that after the Battle of Waterloo in 1815 the military situation had changed and that this might have theoretical ramifications.

The appreciation of the more limited objectives in *contemporary* warfare and a concomitant need for theoretical explanation appear rather late in Clausewitz's work on *On War* itself. Only in book VI does the author start to observe that the great majority of wars and campaigns 'are more a state of observation than a struggle of life and death'.[51] He expresses the suspicion that the Napoleonic wars have proved a mere interlude. While in V.9 he had still

49 Clausewitz, *On War*, 260/Clausewitz, *Vom Kriege*, iv.11, 470; see also Clausewitz, *On War*, 313/Clausewitz, *Vom Kriege*, v.9, 548–9.

50 Clausewitz, *Ueber das Fortschreiten*, 250.

51 Clausewitz, *On War*, 488/Clausewitz, *Vom Kriege*, vi.28, 813.

wondered whether warfare would retain the same levels of energy, he is now ready to answer this question in the negative: 'Not every future war, however, is likely to be of this type; on the contrary, one may predict that most wars will tend to revert to wars of observation'.[52] Hence the realisation that not only a 'true war, or absolute war, if we may call it that', but also the modifications that give it a more limited character deserve full treatment.[53] Two chapters further on he even stresses that any evaluative verdict in which one form of war is placed above the other would be mistaken. One is not more perfect than the other: 'They exist side by side, and their use must be judged on its merits in each individual case'.[54] Yet although the two forms of war are given equal importance, there is still no precise mechanism that explains the possible oscillations between them.

The answer was finally formulated in terms of political objectives which determine the objectives of military action. A notion of the connection between war and politics and the importance of this connection for the difference between absolute and limited war can be found in Clausewitz's earliest work; see for example the following remark in the 'Strategy from the year 1804': 'The political purpose of war can be twofold: to destroy the opponent completely and to end the existence of his state, or to prescribe conditions to him during [the negotiations for] peace'.[55] But he does not take a decisive step until book VIII of *On War*. In VIII.2 he is still groping for a precise explanatory mechanism that relates war to politics.[56] And then within the short space of a mere two pages, at the start of VIII.3B, he formulates the answer by switching on and connecting the interactive and holistic dimensions that had influenced his military thinking all along. He introduces another feedback mechanism (not negative but positive) that is able to catch changes in the military objectives of military opponents (interaction). And he connects this mechanism with the political context of warfare (holism). He expresses an acute interest in the *political* objectives of the parties that wage war, in the different relations between the *states* in question; and in the different stamina of the *governments* in question.[57] The result is *real war* rather than *absolute war*. Real

52 Clausewitz, *On War*, 488/Clausewitz, *Vom Kriege*, vi.28, 813.
53 Clausewitz, *On War*, 488–9/Clausewitz, *Vom Kriege*, vi.28, 813.
54 Clausewitz, *On War*, 516/Clausewitz, *Vom Kriege*, vi.30, 858; see also Kessel, 'Zur Genesis', 412; Gat, *Military Thought*, 224–8.
55 Clausewitz, *Strategy from the year 1804*, 20; see also Paret, *Clausewitz and the State*, 90.
56 Clausewitz, *On War*, 579–80/Clausewitz, *Vom Kriege*, viii.2, 953.
57 Clausewitz, *On War*, 585/Clausewitz, *Vom Kriege*, viii.3B, 960; see also Clausewitz, *On War*, 586/Clausewitz, *Vom Kriege*, viii.3B, 962; Clausewitz, *On War*, 605/Clausewitz, *Vom Kriege*, viii.6B, 990.

war is often limited in the sense that its objectives, its means and its level of violence are limited.[58] Thanks to this insight Clausewitz can now make sense of the fluctuating objectives of the various military conflicts in the history of mankind; these undeniable variations reflect their political objectives.[59] This is his answer to the theoretical riddle that had bemused him already two years after the fall of Napoleon in 'Progression and Suspension'.

This latest stage in Clausewitz's thinking only reaches full maturity in chapter I.1 of *On War*. Right at the beginning of this chapter Clausewitz states his intention to start with the separate elements of his subject, move from there to larger parts and sections and then conclude with the whole in its inner connection. He points out that the subject matter of his book demands this kind of exposition, 'for here more than elsewhere the part and the whole must always be thought of together'.[60] The complexity of war resides in the fact that its parts can only be understood when the whole is grasped first. That is the purpose of the first chapter, aptly titled 'What Is War?' This chapter describes war as an interactive subsystem that is itself shaped in interaction with a wider political system. War is an act of violence whose purpose is to compel the enemy to do our will. This will only happen when the enemy is made utterly defenceless. The interactive process of overcoming the enemy has the form of three explosive positive feedback loops. Each loop only ends when one of the two opponents is defeated. Victory and defeat form the end of military interaction.

The first loop is based on the assumption that in war each side compels the other to do its will. The result is that both sides will raise the application of force to extremes. 'Compelling' implies intention, and intention in its turn is neither exclusively cognitive nor exclusively emotional.[61] This exclusivity is, however, a characteristic of the other two loops.

The second loop is driven by the emotion of fear: as long as I have not overthrown my opponent I must fear that he will overthrow me: 'Thus I am not in control: he dictates to me as much as I dictate to him'.[62] Fear and reactions against it reinforce each other. The result is another dynamic process that tends to extremes.

---

58 Clausewitz, *On War*, 590–1/Clausewitz, *Vom Kriege*, viii.3B, 968–9; see also Clausewitz, *On War*, 606/Clausewitz, *Vom Kriege*, viii.6B, 991; Clausewitz, *On War*, 607/Clausewitz, *Vom Kriege*, viii.6B, 993.

59 Clausewitz, *On War*, 586–94/Clausewitz, *Vom Kriege*, viii.3B, 962–74.

60 Clausewitz, *On War*, 75/Clausewitz, *Vom Kriege*, i.1.1, 191.

61 Clausewitz, *On War*, 75–6/Clausewitz, *Vom Kriege*, i.1.3, 192–3.

62 Clausewitz, *On War*, 77/Clausewitz, *Vom Kriege*, i.1.4, 195.

Finally, the third positive explosive feedback loop is not fuelled by emotion but by a rational activity: the assessment of the power of resistance of the opponent, defined as the product of the means at his disposal and the strength of his will. If I want to beat my enemy I will try to make a calculation of his power of resistance and adjust my own power of resistance accordingly. But my enemy will do exactly the same and the result is again ruthless escalation.[63]

These three abstract escalations will lead us to proclaim 'with inflexible logic' that the greatest possible effort must *always* be exerted in *any* conflict. This means that warfare will move in an absolute direction. In reality, however, the mind refuses to be ruled by such a 'logical fantasy'.[64] If we move 'from the abstract to the real world' we will encounter restrictions that will bend warfare in the direction of a limited rather than an absolute activity. These restraints have again the form of positive feedback mechanisms, but this time the loops are not *explosive* but *implosive*.

The first implosive loop is caused by our evaluation of one of the two factors that form the power of resistance of our opponent: his will (the other factor being the strength of the military means at his disposal). The organisation of our own war effort depends on our previous assessments of our opponent's will. Since human organisation is never perfect, our efforts will fall slightly short of what is needed. This will provide our opponent with a motive to reduce his own exertions, and in that way the perceptions of both parties will have an implosive effect on their mutual will to fight.[65]

Secondly, wars are not resolved at a single blow. Military resources consist of the army itself, but also of its country, its inhabitants, fortresses, and geographic barriers. While the soldiers of an army can be mobilized more or less simultaneously, the other resources can only be used successively. For instance, the use of barriers is determined by the line of advance of the opponent. A similar limitation holds true for allies, who often don't move until the balance of power is threatened. This means that there is always the possibility of making a decision at a later moment. Human reluctance (*Scheu*) to use extreme exertion often makes this the preferred option. Hence Clausewitz's point that 'it is contrary to human nature to make an extreme effort'.[66] In this way the weakness of one party becomes an objective reason for the other party, and vice versa.

63 Clausewitz, *On War*, 77/Clausewitz, *Vom Kriege*, i.1.5, 195; cf. Herberg-Rothe, *Clausewitz's Puzzle*, 46–52.

64 Clausewitz, *On War*, 78/Clausewitz, *Vom Kriege*, i.1.6, 195–6.

65 Clausewitz, *On War*, 78/Clausewitz, *Vom Kriege*, i.1.7, 196–7.

66 Clausewitz, *On War*, 80/Clausewitz, *Vom Kriege*, i.1.8, 199.

The result is again de-escalation, i.e. an implosive positive feedback loop. The will to fight evaporates.

Thirdly, even a lost war frequently is not considered final, and the mere possibility of a more satisfactory outcome in the future will restrain the present exertions of both parties. This constitutes the third and last positive implosive feedback mechanism.[67] This loop very much takes into account the political context of war. The restraint on present *military* exertions is caused by a hope that a remedy may still be found in *political* conditions at some later date.

These three implosive positive feedback loops impart to warfare the character of a probability calculation that includes persons, nations, and relations. At some point in the de-escalating feedback loop, Clausewitz points out, 'the political aim will reassert itself'.[68] This is an intimation of the famous dictum that war is the continuation of politics by other means.[69] But is the link between politics and war always equally strong? Clausewitz's observation that with the three *imploding* loops (i.e. not with the three *exploding* loops) the political objective *reasserts* itself suggests that the relation between politics and war is strongest when military conflict is weakest. Conversely, he notes that in energetic conflicts the military objective takes the place of the political objective, 'as it were' (*gewissermaßen*).[70] And it might seem that only the modifying force of the three implosive loops is able to bring the political objective back to life again. These loops refer explicitly to political actors, i.e. states.[71] Yet if the *explosive* loops that push warfare in an absolute direction seem to blast asunder their political framework, this is only superficially the case. All through the initial sections of chapter I.1 Clausewitz makes a *caveat* (captured by the phrase 'as it were') about the autonomous character of absolute warfare. Later, in section 26 of the same chapter, he stresses that *actually* the political dimension of war is never absent, even during its most acute phases: '[W]hile policy is apparently effaced in one kind of war and yet is evident in the other, both kinds are equally political'.[72] Fluctuations between military objectives are explained by fluctuations in political objectives: 'Thus it follows that without

67 Clausewitz, *On War*, 80/Clausewitz, *Vom Kriege*, i.1.8, 199, 211; see also Aron, *Penser la guerre* 1, 111 and Herberg-Rothe, *Clausewitz's Puzzle*, 52–8.

68 Clausewitz, *On War*, 80/Clausewitz, *Vom Kriege*, i.1.11, 200.

69 Clausewitz, *On War*, 87/Clausewitz, *Vom Kriege*, i.1.24, 210.

70 Clausewitz, *Vom Kriege*, i.1.2, 192; *not* in Clausewitz, *On War*.

71 Clausewitz, *On War*, 81/Clausewitz, *Vom Kriege*, i.1.11, 201.

72 Clausewitz, *On War*, 88/Clausewitz, *Vom Kriege*, i.1.24, 210; see also Clausewitz, *On War*, 87–88/Clausewitz, *Vom Kriege*, i.1.25, 211.

any inconsistency wars can have all degrees of importance and intensity, ranging from a war of extermination down to simple armed observation'.[73]

If Clausewitz developed his third (political) account for constraints on warfare (consisting of positive feedback loops) after his second (suspension) account (consisting of negative loops), then this may signal a lack of satisfaction with the latter account. This would be quite logical; after all, the negative feedback loops of suspension are only able to explain stagnation, whereas the positive loops of the political account are able to actively push levels of violence all the way up or down. Reservations about the explanative power of suspension are indeed already expressed in the very chapter III.16 of *On War* where it is introduced. Immediately after Clausewitz has given his explanations of the suspension of action in war, they receive the following comment: 'Thus, in the midst of the conflict itself, concern, prudence, and fear of excessive risks find reason to assert themselves and to tame the elemental fury of war. But these determinants are hardly adequate explanation for the long periods of inactivity that occurred in earlier wars, in which no vital issues were at stake, and in which nine-tenths of the time that the troops spent under arms was occupied by idleness'.[74] Clausewitz's reservations about the older suspension account may explain why he placed it behind the newer and more potent political account in I.1. The implosive (and explosive) *positive* loops of the political account (I.1.3–11) are able to account for every imaginable decrease or increase in the intensity of warfare, while the *negative* loops of suspension (I.1.12–18) lack this capacity. Moreover, this late change in the order provided Clausewitz with a logical structure, with section I.1.12 acting as a bridge between the two accounts. The political account is able to explain every change up and down the ladder of escalation, but the one thing that it cannot do, as *positive* loops in general cannot, is to provide an account of stasis or balance. Hence Clausewitz's point in I.1.12: 'However modest the political demands may be on either side, however small the means employed, however limited the military objective, can the process of war ever be interrupted, even for a moment? The question reaches deep into the heart of the matter'.[75] Suspension, with the balancing characteristics that are so typical of all *negative* feedback loops, provides a perfectly adequate answer to precisely this deep problem.

73 Clausewitz, *On War*, 81/Clausewitz, *Vom Kriege*, i.1.11, 201.

74 Clausewitz, *On War*, 218/Clausewitz, *Vom Kriege*, iii.19, 409; see also Clausewitz, *On War*, 579/Clausewitz, *Vom Kriege*, viii.2, 952–3.

75 Clausewitz, *On War*, 81–2/Clausewitz, *Vom Kriege*, I.1.12, 201–2.

6 Discussion

In this chapter I have stressed the continuous development in Clausewitz's thinking on limited warfare. The interactive and holistic dimensions, that were present in his earliest thinking, unified his three accounts for the various factors that act as a constraint on warfare. The account of friction starts with the absence of both an interactive and holistic perspective, then interaction is switched on in the account of suspension, and then the holistic perspective is switched on as well in the political account. At the same time these accounts are largely complementary. Although Clausewitz clearly came to favour the political account over the account of suspension, and although this latter account itself had been presented as an improvement on the account of friction, he nowhere suggests that the newer account should actually replace the older account. In chapter III.18 of *On War* he gives as possible causes for suspension both the friction of the first account and the 'new counterweights' (*neu eingetretene Gegenwichte*) of the second accounts.[76] And we have seen that although he expresses reservations about the limited potency of the negative feedback loops of suspension immediately when they are presented in III.16, this account would nevertheless still be included at a much later stage in I.1—but only after he had given the more potent positive loops of the political accounts. The result is a picture of Clausewitz's sophisticated use of causal feedback loops.

My three-stage model provides arguments for continuity where others, including Azar Gat in his impressive *History of Military Thought*, have stressed discontinuity.[77] An important role in Gat's discussion is played by Clausewitz's

76 Clausewitz, *Vom Kriege*, iii.18, 415; 'new counterweights': my own translation; Clausewitz, *On War*, 221 gives 'new opposing forces'; see also the continued use of 'natürliche Friktion' when Clausewitz introduces his account of suspension in *Ueber das Fortschreiten*, 250.

77 My three stage model can also be used as a criterion for completeness. Eugenio Diniz and Domício Proença state that the sole reason why, according to Clausewitz, war never reaches its absolute form is '*because of the superiority of defence alone*' (italics by the authors), i.e. they assume that one of the two explanations given by Clausewitz for suspension forms his sole explanation for limited warfare (Diniz, 'A Criterion', 899).The authors defend Clausewitz's preference for this explanation with a reference to section I.1.17, but I noted above that he only maintains that defence explains '*most* periods of inaction that occur in war', i.e. *not all* periods of inaction (Clausewitz, *On War*, 84/Clausewitz, *Vom Kriege*, i.1.17, 205, italics by me). Already one section and one page later he points out very clearly, in the first sentence (and also in the title): 'There is still another factor that can bring military action to a standstill: imperfect knowledge of the situation' (Clausewitz, *On War*, 84/Clausewitz, *Vom Kriege*, i.1.18, 206). Moreover, Diniz and Proença are correct in observing that this account does not explain the suspension of military activity (which is explained by *negative* feedback loops), but they don't take sufficient note of Clausewitz's

famous Note of 10 July 1827, included by his wife in the first edition of *On War*. In this note he formulated the two novel and interconnected insights that he needed for the formulation of his political account: the double face of war (absolute and limited objectives) and the idea that war is nothing but the continuation of politics with other means.[78] He announced his intention to rework his book in the light of these ideas and hoped that this would impart unity to the entire work. Gat sees the Note from 1827 as a sign of sharp intellectual crisis. This crisis was closely related to Clausewitz's appreciation of the changed political and military situation since the end of the 'heroic period' of the Napoleonic wars in 1815.[79] Gat evokes a dialectical method, in a broadly idealistic sense, as the sudden solution to a sudden crisis about the real or limited versus absolute nature of warfare that can be placed in 1826–1827, i.e. four or five years before Clausewitz's death. The crisis was *sudden* because it occurred only in 1827. The proposed dialectical method was *broad* in the sense that it was not specifically Hegelian and made use of 'theses and antitheses, contradictions, polarity, activity and passivity, positive and negative' that reflected an absorption of the idealist philosophy in general.[80] This method was a *solution* because it helped Clausewitz to conceive absolute and real war as an aspect of a single unity formed by politics. According to Gat, the solution could arise as suddenly as the problem that it solved, because it was only by 1826–1827 that Clausewitz became sufficiently acquainted with idealistic philosophy. Gat notes that Clausewitz's use of a dialectical method started in book VII, increased in book VIII, and reached its peak in the first chapters of book I.[81]

Although Gat has provided a very stimulating reconstruction, two observations can be made. *Firstly*, Hegelian dialectics is a topic fraught with complications; widening this subject to a more general idealistic dialectics carries the risk of making a slippery subject even harder to define and to apply to *On War*. Yet this seems the course taken by Gat himself and by previous commentators.[82] Peter Paret perceived a general dialectical influence, noting that for Clausewitz the use of conceptual contrasts did not imply any 'necessary progression that both gives expression to and moves toward a state of infinite harmony', i.e. he

all-important political account, which is explained by *positive* feedback loops (cf. Diniz, 'A Criterion', 899–90).

78 Clausewitz, *On War*, 69/Clausewitz, *Vom Kriege*, 179–81.

79 Gat, *Military Thought*, 219–20.

80 Gat, *Military Thought*, 232.

81 Gat, *Military Thought*, 232–8.

82 See Gat, *Military Thought*, 233–6; Hahlweg, 'Das Clausewitzbild', 16; Türpe, 'Dialektisches Denken', 709–18; see also Cormier, 'Hegel and Clausewitz', 419–27; Holmes, 'The Clausewitzian fallacy', 1053–6.

plays down a specifically Hegelian influence.[83] Similarly, Raymond Aron was prepared to concede that Clausewitz's method in *On War* is 'dialectical in some way' (*en quelque mésure dialectique*), but he observes preciously few dialectical contradictions that are resolved by a synthesis. Rather than the presence of *contradictions* he notes the use of *contrasts* that evoke 'the polarity of positive and negative electricity'.[84] Indeed, there is little in the general contrasts noted by Paret, Aron and Gat that cannot be explained by the third stage of the model proposed in the present chapter, i.e. Clausewitz's political account.[85] Hence there seems no need for the relatively sudden and yet vague *deus ex machina* of an idealist dialectics.[86] This does not imply a commitment to the strong thesis that *On War* contains no dialectics at all. Rather it prefers a relatively simple account based on the interactive and holistic dimensions that are well-attested in Clausewitz's early and later work, to the sudden inclusion of a problematic concept.[87]

Secondly, and more importantly, Gat's introduction of a sudden dialectical *solution* is formulated as the answer to an equally unexpected intellectual *crisis* that are both dated to 1826–1827. But there is no proof that this crisis ever actually occurred.[88] Clausewitz certainly embarked on a new stage in his thinking about limited warfare with his political account and he certainly intended to revise the older parts of *On War* in the light of this account, as he announced in

83 Paret, *Clausewitz and the State*, 84 n. 13; 150.

84 Aron, *Penser la guerre* 1, 364.

85 See above, §4.5.

86 Cf. Gat, *Military Thought*, 232, 234.

87 But what if the well-established interactive and holistic conceptual dimensions that play such a central role in the present chapter *themselves* were influenced by a broadly idealist context? What springs to mind is the German word for 'interaction', *Wechselwirkung*. This word was used by Immanuel Kant as one of the twelve categories and Clausewitz may have become acquainted with the term through the lectures of Kant's pupil Kiesewetter that he attended. See Kiesewetter, *Versuch einer faßlichen Darstellung*,142–3; Aron, *Penser la guerre* 1, 368–71; see also Gat, *Military Thought*, 177 n. 11. Yet this point is of limited relevance to the present argument, since we have already seen that Gat acknowledges the absence of a 'substantial affinity to the distinctive doctrines of idealism' in Clausewitz's pre-1827 work. Conversely, cf. Aron, *Penser la guerre* 1, 374: 'Tous les germes de ce que l'on peut appeler la dialectique clausewitzienne se trouvent dans les textes antérieurs au *Traité*'.

88 Clausewitz's correspondence in these two years with Gneisenau, who had earlier been the recipient of Clausewitz's accute essay *Ueber das Fortschreiten*, does not show the least sign of intellectual crisis. Actually, this correspondence in 1826–1827 provides us with remarkably few intellectual insights at all. Clausewitz's attention seems limited to such topics as an attack of chest spasms that affected his mother-in-law; see letter from 15 July 1827, in Clausewitz, *Schriften* ii.1, 524.

the Note of 1827. But we have seen that this solution was developed in answer to a problem that he had already perceived on 4 March 1817 at the latest, in 'Progression and Suspension'.

We have seen (§4.5) that Clausewitz had already contemplated warfare with limited objectives as a *contemporary* possibility by 1817—which is what we would expect of a bright observer and participant in a drama that had ended so abruptly in 1815. By that time he was already aware of the theoretical problems caused by these changed historical circumstances. Between 1817 and 1827 he had the time to solve these problems. The year 1827 does not mark a crisis, but rather the formulation (in the form of his political account) of an answer to a ten-year old problem.

What remains is the question why Clausewitz needed so much time to produce an answer. The reply may be that the earlier parts of *On War* were of a more narrowly military nature, while the political dimension of warfare, and hence its matching account gained prominence only at a later stage. This is actually the practical answer that Clausewitz provides himself in the late book VIII on war as an instrument of politics: 'That is why we felt no urge to introduce this point of view at the start. At the stage of detailed study it would not have been much help and might have been distracting. But when plans for a war or a campaign [i.e. book VIII] are under study, this point of view is indispensable'.[89] And although he writes in his Note of 1827 that his new attention to the political context will 'play its part' in his reworking of the first six books, he suggests indeed that this subject is especially relevant to the later books VIII and I.[90]

So, if the dimensions of interaction and holism were indeed relevant at the earliest stages of Clausewitz's work, and if the continuous evolution of his three accounts of the constraints on warfare based on these dimensions is plausible, then there is no need to assume a late and sudden crisis that had to be solved by the equally late and sudden use of a dialectical method.

89 Clausewitz, *On War*, 606/Clausewitz, *Vom Kriege*, viii.6B, 992; see also Clausewitz, *On War*, 605/Clausewitz, *Vom Kriege*, viii.6B, 990.

90 Clausewitz, *On War*, 70/Clausewitz, *Vom Kriege*, 180.

CHAPTER 5

What-if at Waterloo: Clausewitz's Use of Historical Counterfactuals in His History of the Campaign of 1815

1 Introduction

In his *Feldzug von 1815* (*The Campaign of 1815*), Clausewitz makes the following remark about Napoleon after his victory at Ligny against the Prussians on 16 June 1815: 'If Bonaparte had followed with the main army, he would have been ready to fight at Wavre early on the 18th, and it is doubtful whether Blücher would have been in a position to accept a battle at that time and place, and even more doubtful that Wellington could have rushed over in time'.[1] Clausewitz uses a historical counterfactual, which can be defined as a proposition of the form: 'If antecedent A had been the case, consequent C would have been the case'. This counterfactual uses the perfect tense. Strictly speaking, a counterfactual can also be formulated in the imperfect tense: 'If antecedent A *were* the case, consequent C *would be* the case'.[2] In this chapter I will concentrate on counterfactuals in the perfect tense, i.e. I will concentrate on *historical* counterfactuals.

Historical counterfactuals figure in the earliest military historiography and onward; Thucydides and Livy already wondered what their societies would have looked like if the Persians had defeated the Greeks or if Alexander the Great had turned his army toward Rome.[3] Given this omnipresence it would seem that historical counterfactuals have important functions, although debate about the nature of this functionality is by no means resolved, and already

1 Carl von Clausewitz, *The Campaign of 1815* in Christopher Bassford, Daniel Moran, and Gregory W. Pedlow, eds and transl., *On Waterloo: Clausewitz, Wellington and the Campaign of 1815*, (n. p., CLAUSEWITZ.COM, 2010) 186/Carl von Clausewitz, *Feldzug von 1815: Strategische Uebersicht des Feldzugs von 1815*, in *Schriften — Aufsätze — Studien — Briefe. Dokumente aus dem Clausewitz-, Scharnhorst- und Gneisenau- Nachlaß sowie aus öffentlichen und privaten Sammlungen*, ed. Werner Hahlweg, vol. ii.2, §51, 1079. Another translation is Carl von Clausewitz, *On Wellington*, transl. and ed. Peter Hofschröer (Norman, OK: University of Oklahoma Press, 2010).

2 Cf. Lewis, *Counterfactuals*, 1–4; Menzies, 'Counterfactual Theories'.

3 See Rosenfield, 'Why do we ask', 91; Kaye, 'Challenging Certainty', 45–6; Bunzl, 'Counterfactual History', 846.

 | DOI:10.1163/9789004536678_007

Leibniz subjected the concept to trenchant scepticism.[4] Nevertheless, especially since the publication of David Lewis's classical study *Counterfactuals* (1973), large strides have been made toward an increasingly sophisticated philosophical analysis of the functionality of various types of counterfactuals, including historical counterfactuals.[5] These results can be used to obtain a deeper understanding of the functionality of specific types of historical counterfactuals in the works of individual historians, provided that due attention is paid to the historical and historiographical context of their works. In the present chapter I discuss Clausewitz's use of historical counterfactuals in his historical analysis of the Waterloo campaign; I place his preference for *evaluative* counterfactuals in the wider theoretical context provided by *On War*; and I conclude with the relevance of Clausewitz's discussion of counterfactuals for contemporary military historiography.

But why Clausewitz and why the Waterloo campaign? Clausewitz's work on *On War* during his last fifteen years was preceded and accompanied by studies on military history. He took constant care to formulate a general theory of war that remained close to the diversity of historical experience. He was interested in attempts to integrate universality with particularity and was possibly influenced by Montesquieu, either directly, or through his mentor Gerhard von Scharnhorst.[6]

Clausewitz analysed the campaigns of Frederick the Great and other early modern battles, but he was most interested in the more recent battles of the French Revolution and Napoleon, in many of which he had participated himself. His most compelling historical works were written after his appointment as director of the Berlin *Kriegsakademie* in 1818. In 1831 the revolutionary crisis saw him return to active duty in Posen, where he fell victim to the great cholera epidemic one year later. The last work in his hands was probably not the unfinished *On War* but his rather hastily finished history of the French campaigns in Italy and Switzerland of 1799.[7]

The importance of Clausewitz's historical works for the development of his military theory has been widely appreciated, and of these historical works *The*

4 Rosenfield, 'Why do we ask', 90–103; Kaye, 'Challenging Certainty', 42; Evans, *Altered Pasts*, 125.

5 See, among others, Hendrickson, 'Counterfactual Reasoning', 365–86; Menzies, 'Counterfactual Theories'.

6 Gat, *Military Thought*, 169, 191, 195. See above, §3.2, for Montesquieu and the historical meso-level as the locus where micro-level and macro-level are combined.

7 Kessel, 'Zur Genesis', 418–23; Paret, *Clausewitz and the State*, 328–30; Hahlweg, preface in Clausewitz, *Feldzug*, vol. ii.2, 936–7.

Campaign of 1815 ranks especially high.[8] It was written between 1827 and 1830 and he used its material to teach Crown Prince Friedrich Wilhelm of Prussia.[9] While Clausewitz had still treated historical description and critical analysis separately in his account of the campaign of France of 1814, his account of the military events of 1815 achieves an exemplary and seamless integration of description and analysis in 57 short sections. Possibly there was a direct connection between Clausewitz's analysis of Waterloo and his mature appreciation of the relation between war and politics.[10] Indeed, *The Campaign of 1815* contains many concepts that are developed at a more abstract level in *On War*. This holds true for the ideas of friction and the culmination point of an attack; and, most significantly, the same point can also be made for historical counterfactuals.[11]

Moreover, the Battle of Waterloo itself made a career as the historical counterfactual *par excellence* from the moment when Arthur Wellesley, Duke of Wellington, famously observed that the outcome had been 'the nearest run thing you ever saw in your life', until the fatal shots fired at Sarajevo on 28 June 1914 provided the world with a more recent counterfactual paradigm.[12] The counterfactual of a French victory at Waterloo was used in 1907 by G.M. Trevelyan and other authors as the starting point for a full-fledged 'alternate' history.[13] Finally, the Waterloo campaign sees a remarkable combination of the same persons in multiple 'counterfactual roles'. Napoleon was not only the object of many counterfactuals, but also used them avidly himself when, in the deadly torpor of St. Helena, he dictated *Mémoires* that were specially dedicated to the Waterloo campaign.[14] Clausewitz himself participated in the campaign as chief of staff of the Prussian corps that pinned down Marshal Grouchy at Wavre.[15] In addition, he wrote the counterfactuals in his own account often in a direct reaction against those of Napoleon, whose *Mémoires* in many ways set a

---

8 Daniel Moran, 'Clausewitz on Waterloo'; Paret, *Clausewitz and the State*, 329; Sumida, *Decoding Clausewitz*, 136; Colson, 'Clausewitz on Waterloo', 440; Herberg-Rothe, *Clausewitz's Puzzle*, 32–8, 79–85.

9 Bassford, introduction in Clausewitz, *Campaign*, 6; Moran, 'Clausewitz on Waterloo', 237. Herberg-Rothe, *Clausewitz's Puzzle*, 37, dates the writing of the *Campaign of 1815* to 1827–1828.

10 Herberg-Rothe, *Clausewitz's Puzzle*, 83–4.

11 Friction: Clausewitz, *Campaign*, 123–4/Clausewitz, *Feldzug* §33, 1013–14; culmination: Clausewitz, *Campaign*, 122–3/Clausewitz, *Feldzug*, §33, 1012. On friction see also above, §4.3; on culmination see §4.4.

12 Quoted in Moran, 'Clausewitz on Waterloo', 253.

13 Trevelyan, 'If Napoleon Had Won', 184–200; see also Uffindell, 'Napoleon at Waterloo', 187–202; Tsouras, *Napoleon Victorious*; Kaye, 'Challenging Certainty', 48.

14 Napoléon, *Mémoires*; see also Evans, *Altered Pasts*, 3; Moran, 'Clausewitz on Waterloo', 253.

15 See Stoker, *Clausewitz*, 223–53.

polemic agenda for Clausewitz. Finally, Wellington was one of the main actors in the military drama, the recipient of some major criticism from Clausewitz, and also the writer of a reaction against this criticism, in which he scornfully rejects the latter's frequent use of counterfactuals—immediately after having employed an extensive counterfactual himself.[16]

2 Description: the Waterloo Campaign in 25 Counterfactuals

Clausewitz starts his account of the extraordinary events that saw Napoleon escape from his exile on Elba and march across France, on 20 March 1815—the day Napoleon triumphantly entered Paris and King Louis XVIII took refuge to Ghent. Napoleon started the 'Hundred Days' with energetic attempts at internal pacification and with preparations for the inevitable clash with the coalition of four great powers that had defeated him and forced him to abdication in the previous year. An Anglo-Dutch and a Prussian army would enter the southern part of the Kingdom of the Netherlands soon, and Austrian and Russian armies were on their way to other parts of the French frontier (the Seventh Coalition). Clausewitz concentrates on Napoleon's military preparations and the subsequent Waterloo campaign, but he never loses sight of the precarious political context for the Emperor's last military adventure. The *Campaign of 1815* is positively bristling with historical counterfactuals. It is possible to summarize Clausewitz's history of the campaign in the form of a series of counterfactuals, which is what I shall do here, with each counterfactual preceded by a number between square brackets.

[1] If Napoleon had imposed general conscription for all male French citizens between the ages of 20 and 60, he would have had a considerable force of over 2,000,000 soldiers. [2] Yet, this measure would only have worked if he had sufficient equipment and if the French had stood unified and enthusiastically behind him. His own *Mémoires*, as Clausewitz is keen to point out, show that he entertained grave doubts on each of these conditions; hence he rejected this option.[17] Napoleon also considered the possibility of a defensive

16 Wellington, *Memorandum*, 227: 'It is useless to speculate upon supposed military movements which were never made, and operations which never took place, or the object of the several chiefs of Generals opposed to each other'. Wellington, 'Memorandum', 226: 'But let us consider whether the abandonment of all the objects which the Allies had in view in maintaining any position in the Netherlands would have enabled the Generals of the Allied armies the better to fight a great battle with the enemy'.

17 Clausewitz, *Campaign*, 63–4/Clausewitz, *Feldzug*, §5, 953; Napoléon, *Mémoires*, 51–65.

strategy that would have seen France invaded by the forces of the Coalition, but would have given him extra time to reinforce, make use of French fortifications and possibly incite his citizens to a patriotic insurrection. [3] But a foreign invasion would have further compounded his already extremely perilous internal political situation. His *Mémoires* clearly testify that he felt this danger through and through. Hence, it is not surprising that he rejected this option as well.[18] So, Napoleon had no other option than an offensive operation with an army that numbered 129,000 men against his nearest enemies, the 99,000 soldiers of the Anglo-Dutch under Wellington and the 115,000 Prussians under Gebhard Leberecht von Blücher—in that sense he took the right decision. [4] Yet his chances of success were exceedingly slim and even a resounding blow against the English and Prussians would still have left the possibility of defeat at the hands of the Austrians and the Russians. [5] But if Napoleon had managed to crush Wellington and Blücher, he would have been able to spare troops to counter the Austrians and the Russians while drawing on the vast capabilities of an energized and reunited France. The overall odds would still remain against him, but this would have given him a real chance to take on the Coalition.[19] Given Napoleon's urgent need for success against both the Anglo-Dutch and the Prussians, and given the well-known energetic way in which he waged his campaigns, Wellington and Blücher should have tried harder to combine their forces. [6] Given his desperate need for a resounding victory, Napoleon would have attacked them anyway, even if he had been forced to confront them jointly. [7] But even if it had been obvious to both allied commanders that such a combination of forces was possible, it would still not have happened because Wellington was unduly preoccupied with the continued possession of Brussels. This fixation prompted him to deploy his troops in a wide area south of the city, screening its multiple approaches.[20]

On the night of 14–15 June the Coalition forces were surprised to learn that Napoleon was on the move and poised to strike in the direction of Charleroi, which was taken on the fifteenth. Napoleon had managed to insert himself between the Anglo-Dutch in the northwest, and the Prussians in the northeast. On the 16th he sent Marshal Ney with 48,000 men to take and guard Quatre Bras against the Anglo-Dutch and decided to use his remaining 75,000 men against the Prussians, whom he attacked and beat, but did not destroy, at Ligny that same day (see Figure 2). [8] The Prussian General Bülow had remained too long in Liège, although he had received timely marching orders. Hence

18 Clausewitz, *Campaign*, 66–8/Clausewitz, *Feldzug*, §7, 956–8; Napoléon, *Mémoires*, 59–60.
19 Clausewitz, *Campaign*, 180–81/Clausewitz, *Feldzug*, § 14, 971–72.
20 Clausewitz, *Campaign*, 183–84/Clausewitz, *Feldzug*, §15, 974.

he would have been able to reach Ligny twelve hours earlier than he actually did, just in time to decide the battle in favour of the Prussians.[21] [9] Similarly, Wellington had been able to send reinforcements that could have reached Ligny in time, again deciding the battle in favour of the Coalition, but he continued his passive defence of the approaches to Brussels.[22] [10] On the French side, Napoleon maintained in his *Mémoires* that he had ordered Ney to send 10,000 men from Quatre Bras to Ligny; in that way the Prussians would have been surrounded, and Napoleon boasted that they would have been taken *en flagrant délit*.[23] [11] Clausewitz does not agree: if Napoleon had really had this aim, he would have given clear and timely orders, but proof for such orders is conspicuously absent.[24] [12] And even if Ney had appeared with 10,000 men in the open territory around Ligny, then Blücher would probably have been able to free part of his 80,000 men for countermeasures.[25] [13] Clausewitz also notes that if Ney had possessed complete knowledge of the situation, he would have realized that he actually was able to drive away the Anglo-Dutch from Quatre-Bras and then fall on the right wing of the Prussians at Ligny in full force, all on 16 June.[26]

After the Battle of Ligny, Blücher's forces were battered but still intact. He started a difficult retreat in the northern direction of Wavre with the aim of joining forces with Wellington. Napoleon maintains in his *Mémoires* that he was aware of this manoeuvre, but Clausewitz retorts that [14] if he had indeed guessed this movement, he would not have sent merely 35,000 men under Marshal Grouchy to go after the Prussians.[27] [15] Actually, if Napoleon had followed the Prussians swiftly and with all his forces, and forced them to another battle, it is doubtful whether they would have been able to survive this second onslaught, and even more doubtful whether the Anglo-Dutch would have arrived in time to save their allies. In that way, Napoleon could have decided the entire campaign.[28] [16] But even without such a major shift in Napoleon's operations, it still remains true that if he really had guessed the direction

21 Clausewitz, *Campaign*, 90/Clausewitz, *Feldzug*, §20, 980.
22 Clausewitz, *Campaign*, 90–1/Clausewitz, *Feldzug*, §21, 980–1.
23 Napoléon, *Mémoires*, 93–4, 100; Clausewitz, *Campaign*, 106–7/Clausewitz, *Feldzug*, §30, 996.
24 Clausewitz, *Campaign*, 107–11/Clausewitz, *Feldzug*, §31, 997–1000; Clausewitz, *Campaign*, 131–2/Clausewitz, *Feldzug*, §36, 1023.
25 Clausewitz, *Campaign*, 111–12/Clausewitz, *Feldzug*, §31, 1001.
26 Clausewitz, *Campaign*, 131–2/Clausewitz, *Feldzug*, §36, 1023–4.
27 Napoléon, *Mémoires*, 115–17; Clausewitz, *Campaign*, 134–5/Clausewitz, *Feldzug*, §37, 1025–26.
28 Clausewitz, *Campaign*, 185–187/Clausewitz, *Feldzug*, §51, 1079–1081.

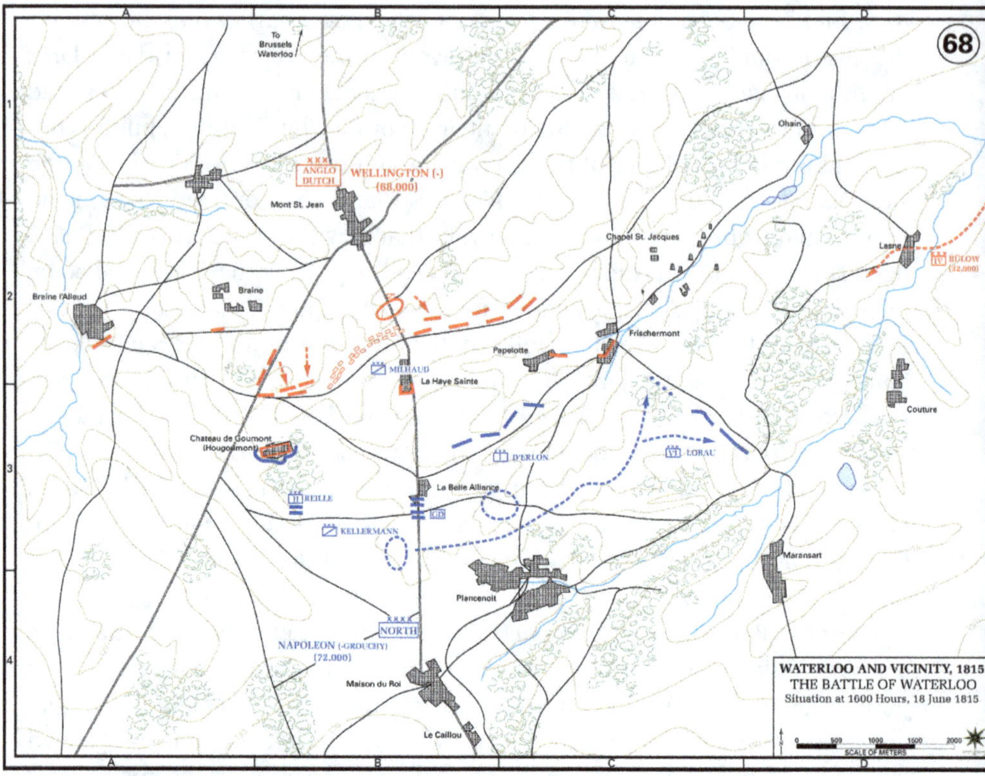

FIGURE 2 The battles of Ligny and Quatre-Bras, 16 June 1815
SOURCE: HTTPS://WWW.WESTPOINT.EDU/SITES/DEFAULT/FILES/INLINE-IMAGES/ACADEMICS/ACADEMIC_DEPARTMENTS/HISTORY/NAPOLEONIC%20WARS/NAPSEL119.GIF, ACCESSED 28 OCTOBER 2020. COURTESY OF THE UNITED STATES MILITARY ACADEMY DEPARTMENT OF HISTORY

of Blücher's retreat, it would have been more natural (*natürlicher*) to post a strong corps on the left bank of the Dyle. [17] If a corps had been placed there, it would have been able to prevent the Prussians from joining the Anglo-Dutch at Waterloo while at the same time being available to help Napoleon in his decisive battle against the Anglo-Dutch.[29] Napoleon would later maintain that it had been his plan all along for Grouchy to come to his aid against the Anglo-Dutch, and that [18] if Grouchy had followed his orders correctly, he would have arrived in time at the battlefield of Waterloo, thus deciding the battle in favour of the French.[30] Clausewitz remarks that there is again no proof in the form of any clear order for this, and [19] if this had indeed been Napoleon's

29 Clausewitz, *Campaign*, 134–35/Clausewitz, *Feldzug*, §37, 1026.

30 Napoléon, *Mémoires*, 107–15, 142–58, 197.

intention, he should have taken care from the start to keep Grouchy *between* himself and the Prussians, not *behind* the Prussians.[31] In the event, Grouchy first missed the Prussians and then ended up with a battle against their rear guard on 18 June at Wavre, while their regrouped and replenished main force was able to partake in the battle against Napoleon at Waterloo that same day.

Meanwhile, Napoleon had used 17 June to push on in the direction of Wellington, and 18 June saw his army near Waterloo, pitched against the Anglo-Dutch, who later that day would be reinforced by the Prussians (see Figure 3). Napoleon opened his attack as late as 2:00 p.m. He could have attacked between 6:00 and 7:00 a.m. Why did he think that he could afford to wait? Because he did not believe that Wellington would be able to bring in any more additional troops, and also because he did not expect Blücher to come to Wellington's aid that afternoon. Napoleon was right in the first assumption and his miscalculation in the second assumption did not matter very much; [20] if he had attacked earlier, then Blücher in his turn probably would have found ways to appear earlier as well, so that it would still have been hard for Napoleon to win at Waterloo. So neither assumption had much impact on the final outcome of the battle. Yet Napoleon's decision should be judged on the basis of his assumptions at that moment. His arguments for these assumptions were defective in both cases.[32]

The French concentrated their attack during the Battle of Waterloo on Wellington's *centre,* which however they did not manage to break. [21] If they had rather concentrated on Wellington's *left wing,* which was relatively weak and exposed by open terrain, the chances of success would have seemed higher.[33] [22] But this would probably have exposed the French rear to an attack by a sizeable detachment of Prussian reinforcements. So an attack against Wellington's left wing would actually have been the worst rather than the best option. [23] If Napoleon had included the timely arrival of the Prussians on the battlefield in his plans, as indeed he should have, then he should have tried an attack on the Anglo-Dutch *right wing*, even though this wing profited from depressions in the landscape.[34] [24] In that case the Anglo-Dutch would still have been supported by the Prussians, and an Anglo-Dutch rout would still have been unlikely; but perhaps the combined coalition forces might have suffered a similar blow as the Prussians had suffered two days earlier at Ligny; perhaps this setback would have caused hesitation and discord in the allied

31 Clausewitz, *Campaign*, 160–67/Clausewitz, *Feldzug*, §48, 1053–59.
32 Clausewitz, *Campaign*, 160/Clausewitz, *Feldzug*, §48, 1052.
33 Clausewitz, *Campaign*, 174/Clausewitz, *Feldzug*, §48, 1066.
34 Clausewitz, *Campaign*, 174–5/Clausewitz, *Feldzug*, §48, 1067.

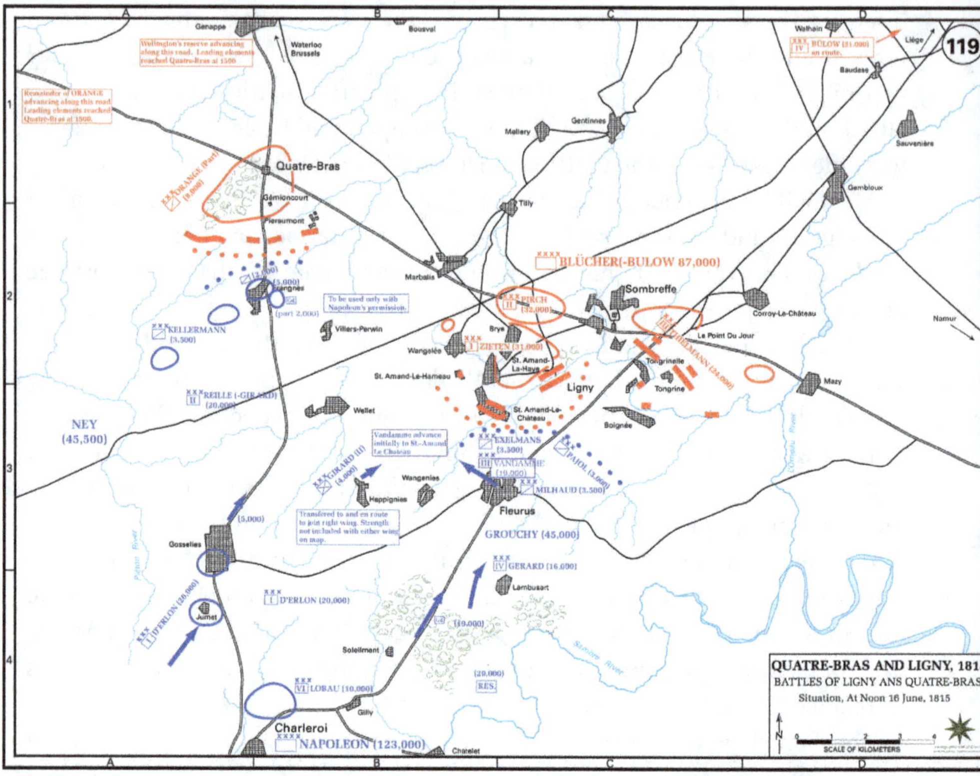

FIGURE 3 The Battle of Waterloo, 18 June 1815
SOURCE: HTTPS://WWW.WESTPOINT.EDU/SITES/DEFAULT/FILES/INLINE-IMAGES/ACADEMICS/ACADEMIC_DEPARTMENTS/HISTORY/NAPOLEONIC%20WARS/NAP68.PDF, ACCESSED 28 OCTOBER 2020. COURTESY OF THE UNITED STATES MILITARY ACADEMY DEPARTMENT OF HISTORY

command; perhaps in this constellation Grouchy's arrival a day later might have had an additional effect; and perhaps this would have marked the beginning of more sizeable results for the French.[35] [25] As it was, the fatal outcome of Napoleon's late and uncoordinated attack against the Anglo-Dutch centre during the Battle of Waterloo would probably not have been reversed even if Grouchy had arrived in time on the battlefield of Waterloo. Since the Prussians would have managed to send in additional troops as well, the cataclysm for the French would have been all the more comprehensive.[36] Clausewitz notes with grim satisfaction that this time it was the turn of 'the great magician' himself to be taken *en flagrant délit*—not in a historical counterfactual, but for real.[37]

35 Clausewitz, *Campaign*, 175/Clausewitz, *Feldzug*, §48, 1067.
36 Clausewitz, *Campaign*, 180–1/Clausewitz, *Feldzug*, §50, 1075.
37 Clausewitz, *Campaign*, 191/Clausewitz, *Feldzug*, §50, 1087.

3 Analysis of the Function of Counterfactuals in the *Campaign of 1815*

In his critical analysis of the Battle of Ligny, Clausewitz excuses the modest performance of the Prussians and their commanders with the remark that action in war is like moving in a resistant element, so that even the most mediocre results demand uncommon capabilities. This remark about the debilitating effect of friction is used to emphasize the function of criticism: it should try to assess the truth, not exercise the function of a judge.[38] Peter Paret indeed used this observation to stress that Clausewitz 'considered judgment in the sense of approval and disapproval to be ... infinitely less significant than understanding what had occurred'.[39] A good case can indeed be made for the Clausewitz's disinterested views on political systems and ideologies.[40] When it comes to individuals, however, Clausewitz was very much interested in passing judgment. He constantly hands out praise and blame, implicitly or explicitly. Keywords that appear again and again in this context are 'to forgive' (*verzeihen*) and, especially, 'to blame' (*tadeln*).[41] The *Campaign of 1815* was composed by an officer who participated in the events himself and who evaluates the performance of his fellow officers. For instance, he notes that the Coalition was still in the dark about Napoleon's whereabouts until 14 June and that it remained too long in a condition of 'blameworthy indecision' (*tadelnswerther Schwebe*). The Coalition had an—admittedly—vague sense of danger and yet it managed to be completely surprised by the lightning action of the Emperor.[42] And when Clausewitz discusses Napoleon's decision to continue the Battle of Waterloo in spite of the arrival of the Prussians on the battlefield he adds that he cannot be blamed. Given his extremely vulnerable political internal and international situation, he had to persevere against all odds—it is not without some admiration that Clausewitz adds: 'There are situations when the greatest prudence can only be sought in the greatest boldness, and Bonaparte's situation was one of them'.[43] Since the expression of approval and disapproval is so important for

38 Clausewitz, *Campaign*, 123–124/Clausewitz, *Feldzug*, §33, 1014; cf. Clausewitz, *On War*, 120/Clausewitz, *Vom Kriege*, ii.7, 263; see also above §4.3.

39 Paret, *Clausewitz and the State*, 354.

40 Paret, *Clausewitz and the State*, 349.

41 Girard, *Battling to the End*, 147.

42 Clausewitz, *Campaign*, 87/Clausewitz, *Feldzug*, §18, 978.

43 Clausewitz, *Campaign*, 177/Clausewitz, *Feldzug*, §48, 1070.

Clausewitz, it is not surprising that a large part of his counterfactuals can be understood in this light.[44]

When Clausewitz uses *evaluative* counterfactuals his aim is frequently the assessment of responsibility, often in a direct reaction against Napoleon himself. In his *Mémoires* the Emperor constantly tried to shift the blame for the indecisive nature of the Battle at Ligny to Ney [10] and the disaster at Waterloo to Grouchy [18] who, Napoleon maintains, both failed to follow his orders. Clausewitz counters these counterfactual arguments in two different ways.

Firstly, he notes that both are based on the assumption that Napoleon issued the orders in the first place; but Clausewitz denies that these orders were given, and without proof of their existence the ramifications of their non-execution become vacuous. Moreover, in both cases Clausewitz formulates his refutation of Napoleon's counterfactual in the form of new counterfactuals, so that [11] reacts against [10] and [19] reacts against [18]. Counterfactual [19] is especially interesting: if Napoleon had indeed intended to use Grouchy's help at Waterloo before the arrival of the Prussians, then he should have taken care to keep Grouchy between himself and the Prussians. Usually both the antecedent and the consequent of a counterfactual take the form of *a false event* (i.e. an event that did not take place), but in this case only the antecedent is a (false) event. The consequent is not factual but deontological; 'Napoleon *should* have'.

Clausewitz's second mode of attack is to allow for a moment the truth of Napoleon's counterfactuals, but to deny their relevance and hence their success in shifting the blame away from the Emperor to his subordinates. The effect of Clausewitz's argument is again exculpation of Ney [12] and Grouchy [25], and hence inculpation of Napoleon. Even if his supposed orders had been obeyed, the result would still have been unsatisfactory, given the impact of his other, more fundamental errors. Both counterfactuals have again an intriguing form. Rather than the usual case of an antecedent and a consequent that are both false, in these cases we have an antecedent fact that is indeed false, but a consequent that is not false but true: 'Even if Grouchy had arrived in good time on the battlefield (not a fact), Napoleon would still have lost at Waterloo (a fact)'. Nelson Goodman has called these partial counterfactuals *semifactuals*, which typically have the form 'even if ... then still' rather than 'if ... then'.[45]

This second mode of attack on Napoleon's attempts at self-justification points to the importance of the (un)availability of alternative options and

44 See the counterfactuals [2], [3], [6], [7], [8], [11], [12], [14], [16], [17], [18], [19], [20], [21], [23] [24], and [25].

45 Goodman, *Fact*, 3–31. Other semifactuals employed by Clausewitz include [4], [6], [7] and [25].

outcomes to an historical actor. These factors are highly relevant if we want to judge the person in question. Clausewitz seems to subscribe to the common assumption that the potential for praise and blame will be higher to the extent that an actor had more options that were able to produce more different outcomes. Since he adheres to this assumption about evaluative verdicts in general, it is not surprising that this assumption also informs his use of evaluative counterfactuals. Blame is often assigned in the form of a counterfactual that points to the availability of a different option that could have led to better outcomes. Wellington would have been able to send reinforcements to the Battle at Ligny and this could have decided the fight in favour of the Coalition, but he took the culpable decision to remain passive [9].[46] On the other hand, Napoleon's options after his return from Elba were very limited. The chances of success of an *offensive* military campaign against the assembling Seventh Coalition were very small, and even complete success against the Anglo-Dutch and the Prussians would still have left the gathering Russians and Austrians to be dealt with [4]; but a *defensive* campaign would have been fraught with (political) dangers as well [3]. The combination of [3] and [4] amounts to a semifactual: the *alternative* antecedent (a defensive versus an offensive campaign) is given the same likely consequent (a defeat for Napoleon) as the *actual* antecedent. This semifactual is used to excuse Napoleon's actual choice.

But this is merely a simplified sketch of Clausewitz's approach. He realizes very well that a fair counterfactual evaluation of his actors should not be based exclusively or even primarily on their objective options and their related outcomes, but rather on their own *perceptions and evaluations*. He tries to present an analysis of the aims and means of the main historical actors that is based on their own situational knowledge. Only with this information does he feel safe to pronounce evaluative judgments. And here again, historical counterfactuals are used to elucidate the options of the historical actors in question. At the start of the Hundred Days, Napoleon considered the option of a general conscription. This would have provided him with a very large army of some 2,000,000 recruits ([1]); but he rejected this option because of its expected negative outcome. Political dissent stood in the way of raising this army and the required equipment was lacking ([2]). Napoleon 'most definitely' (*auf das allerbestimteste*) felt the problems related to this option; his analysis was correct, and hence he is not blamed for rejecting this option.[47] Similarly, on 16 June Marshal Ney had the means to drive away the Anglo-Dutch from

46 See also counterfactual [8].

47 Clausewitz, *Campaign*, 63/Clausewitz, *Feldzug*, §5, 953.

Quatre Bras and subsequently fall on the right wing of the Prussians at Ligny [13]; but because he did not and could not have the situational knowledge to see this as a viable option, he receives no blame.[48]

So far, we have seen that Clausewitz uses counterfactuals for the attribution of responsibility, often in a polemical context, for the assessment of available options and outcomes, and for a better understanding of the situational knowledge, the aims, and the means of his historical actors. All these points are related to the central function of administering praise and blame, while this activity itself is again frequently expressed in the form of counterfactuals. In the next section we will see how this use of evaluative counterfactuals fits the wider didactic context of Clausewitz's historical criticism as explained in *On War*.

Let us first see, however, how the case of Clausewitz relates to a modern account of the function of historical counterfactuals in historiography and what the differences may teach us. Daniel Nolan lists eight functions; he notes that counterfactuals are used because: (1) they can invigorate the historical imagination and suggest new hypotheses for investigation; (2) they help in bringing out disagreement by making underlying assumptions about the course of events clearer; (3) they mitigate hindsight bias and increase the appreciation of historical contingency; (4) we gain inside understanding when we partake in the counterfactual worries and assessments of historical actors; (5) people are curious about counterfactual questions, which seems to make them legitimate topics of inquiry in their own right; (6) historians make causal judgments and (7) they give explanations, and the use of counterfactuals is closely related to these activities; and (8) counterfactuals can inform the value judgments of historians, including the assessment of responsibility.[49] It is easy to find numerous examples of these points in the *Campaign of 1815*, but I will concentrate on two issues.

The most contentious issue is Nolan's third point: the mitigation of hindsight bias and the increased appreciation of historical contingency. This might imply that counterfactuals can be used to make a point about the contingent versus deterministic character of history in general. This function has indeed been ascribed to counterfactuals by Niall Ferguson and other modern historians. Ferguson gives a brief history of historical determinism, presents contingent or 'chaotic' history as an alternative, and argues for the anti-determinist function of counterfactuals.[50] But sometimes a historical consequent is the

48 See also Paret, *Clausewitz and the State*, 332; Moran, 'Clausewitz on Waterloo', 240.

49 Nolan, 'Why historians should care', 333–4.

50 Ferguson, introduction in *Virtual History*, 20–88; see also Kaye 'Challenging Certainty', 38–43.

very contrary of contingent; it is overdetermined. A consequent is overdetermined when it has multiple antecedents that can each be regarded as a cause, while already a mere subset of these causes would have been enough to trigger the consequent. It has been argued that counterfactuals can perform an important function in solving problems of overdetermination. They can help us to make a distinction between antecedents that are sufficiently potent to trigger a consequent and antecedents that lack this capacity; in this way, historians use counterfactuals to assess 'difference makers': 'If antecedent A had not been the case, consequent C would not have been the case'.[51] One conclusion can be that an antecedent A, or even a whole series of antecedents, would *not* have made a difference. In this way a counterfactual can be just as easily used to make an 'overdetermined' point as a contingent point.[52] In Clausewitz we can indeed observe a perfect agnosticism about the contingent or determinist character of history as such.[53] He sometimes observes that many options were open to the historical actors in question, and sometimes that there was an acute paucity of such options. Sometimes history is very open and sometimes it is very closed.[54] We have seen that this position by no means impedes a fruitful use of counterfactuals. Actually, one of Clausewitz's major reasons for using counterfactuals in the first place, is to understand or to assess *how* open or closed a given case actually was; and if he makes a point about the closed character of a case, he often uses a semifactual: 'even if ... then still'.

While Clausewitz thus (quite correctly) gives less prominence to the third point in Nolan's list of counterfactual functions (appreciation of contingency), the opposite holds for Nolan's eighth point: the use of counterfactuals in making value judgments, including the assessment of responsibility. It is interesting to note that this function figures as the last item on Nolan's modern list, and this is not coincidental. Nolan writes: 'One use of counterfactuals for historical purposes that has rarely been focused on in the recent literature is in the attribution of responsibility, and in the determination of the appropriateness of regret and pride, and to a lesser extent praise and blame'.[55] He notes that when it comes to the attribution of responsibility and determination of the appropriateness of regret and pride, we are especially interested in whether

51 Cf. Bunzl, 'Counterfactual History', 857; Reis, 'Counterfactuals', 722.

52 Kaye, 'Challenging Certainty', 50; Evens, *Altered Pasts*, 36–7, 61.

53 See also Moran, 'Clausewitz on Waterloo', 246.

54 See also Montesquieu's aversity from an priori commitment to historical determinism in *De la politique* (1725) and in the *Reflexions sur la charactère de quelques princes* (1731–1733), above, §3.5.

55 Nolan, 'Why historians should care', 331.

the outcome of the event would have occurred anyway. When the outcome is inevitable, i.e. when we start with a counterfactual question that is given a semifactual answer, we will feel that this reduces the responsibility of any specific individual for that outcome. We have already seen how, according to Clausewitz, the responsibility of Napoleon for an offensive campaign against the Seventh Coalition was reduced by the inevitable character of an offensive versus a defensive campaign.[56] Insofar as many modern historians have turned away from individuals toward impersonal processes and structures, it is not surprising that the use of counterfactuals for the attribution of personal responsibility has not received much attention. Yet in spite of this apparent modern rarity, we have seen how evaluative counterfactuals play an essential role in Clausewitz's *Campaign of 1815*. Moreover, while Nolan merely aims to present a list of different and only loosely connected counterfactual functions, Clausewitz's account shows how different counterfactual functions can form an intricate and cohering pattern in which a central position is taken by evaluative judgments.

Clausewitz's use of counterfactuals for an evaluation of the situational knowledge of historical actors and his consistent use of personal praise and blame in his *Campaign of 1815* point to a great interest in what Daniel Moran describes as 'the minds of the men who commanded the armies that fought [this campaign]'.[57] In no mind is Clausewitz more interested than in Napoleon's and he constantly reacts against Napoleon as actor and as writer of the *Mémoires*. Napoleon is the object of intense criticism—although Blücher and, especially, Wellington are not spared either.[58] With his failure to pursue the Prussians after his victory at Ligny, Napoleon squandered his best chance at a decisive victory (see [15]). This behaviour is compared with two cases in previous campaigns. These cases are formulated in similar counterfactual terms and contribute to the construction of a pattern. If Napoleon had chased the Austrians after the Battle of Dresden in 1813, and the Prussians after they had been caught at the Marne in 1814, he would have been able to follow up partial success with decisive victory. In his later years he had become used to military opponents that took flight or became paralyzed after the first blow. He had formed a habit of underestimating and disdaining his adversaries that contributed to his final undoing in 1815.[59] Yet even here, where Clausewitz's criticism

56 See above §5.3.

57 Moran, 'Clausewitz on Waterloo', 240.

58 Cf. Hahlweg, preface in Clausewitz, *Feldzug*, 939.

59 Clausewitz, *Campaign*, 63/Clausewitz, *Feldzug*, §51, 1080; see also Clausewitz, *On War*, 162/Clausewitz, *Vom Kriege*, ii.5, 322–323.

of Napoleon is extremely harsh, his fascination for Napoleon reaches levels where it takes the form of downright identification. The use of counterfactuals allows Clausewitz to step into the shoes of Napoleon and show how his mistakes could have been prevented. Take the point about Napoleon failing to follow the Prussians after they had been caught at the Marne in 1814 and after they were defeated at Ligny. Clausewitz's point in both cases is that Napoleon failed because he was not decisive enough against the Prussians. Clausewitz seems to use evaluative counterfactuals to show how *he himself* would have succeeded by acting more Napoleonic than Napoleon himself.[60]

At the same time Clausewitz uses Napoleon's own *Mémoires* as ammunition for an uncanny picture of gradual mental decline with disastrous consequences. He uses Napoleon's efforts to shift blame to Ney on 16 June (see [10]) and to Grouchy on 17 June (see [18]) to corroborate a pattern of vague and hesitant commands: 'Given all this, it can already be said that even on the 16th Bonaparte was no longer equal to the task that fate had imposed upon him'.[61] This line of argument is continued on the day of the Battle of Waterloo on 18 June. Napoleon loses many hours deploying his army in a parade-like formation. He enjoys the accolades of his troops while riding in front of them. According to Clausewitz, by this time the Emperor seems to have lost faith in his own campaign. With 'the unnecessary assembling and parading of his army' he hoped to induce Wellington's army to retreat so that, in a striking departure from his previous habits, he could evade battle altogether.[62] All this confirms for Clausewitz the impression 'that something had changed within him'.[63] This mental unravelling contributed to Napoleon's futile attack on the Anglo-Dutch centre. Due to the unexpected arrival of the Prussians, he must have realized that his attack had little chance of success; his decision to start this attack was an act of 'sheer desperation'.[64] His decision to continue this attack while defeat was already staring him in the face lacked all rationality and no longer showed Napoleon as a great man, but 'like someone who has broken an instrument and in his anger smashes the parts to pieces on the ground'.[65]

Although Clausewitz's evaluative counterfactuals in the *Campaign of 1815* are used to establish a pattern of personal hubris and failing leadership, this

60 Girard, *Battling to the End*, 137–155.

61 Clausewitz, *Campaign*, 127/Clausewitz, *Feldzug*, §34, 1018.

62 Clausewitz, *Campaign*, 160/Clausewitz, *Feldzug*, §48, 1052–3.

63 Clausewitz, *Campaign*, 167/Clausewitz, *Feldzug*, §48, 1058.

64 Clausewitz, *Campaign*, 173/Clausewitz, *Feldzug*, §48, 1065.

65 Clausewitz, *Campaign*, 178/Clausewitz, *Feldzug*, §48, 1071; cf. Mongin, 'A Game-Theoretic Analysis'.

amounts to more than just a case of old-fashioned nineteenth-century 'Great Man' historiography. In order to appreciate this point we need to turn to Clausewitz's theoretical masterwork.[66]

4 Context: on War

In *On War* Clausewitz explains that the theory of war concerns itself with the study of ends and means.[67] Various factors conspire to make this study a highly problematic venture: war is driven by complex psychological forces; these forces tend to interact; and in general all military action has to proceed in a state of twilight due to a constant lack of information.[68] Nevertheless it is possible to formulate a large number of evident propositions, for instance that defence is the stronger form of warfare with a negative purpose while attack is the weaker form with a positive purpose; and that victory not only implies the conquest of the battle field but also the physical and psychological destruction of the opponent.[69] History is an important tool in the formulation of these rules, if only because it provides certain limiting conditions: 'theoretical results must have been derived from military history or at least checked against it'.[70] Neither here nor elsewhere does Clausewitz provide a precise methodological procedure for the contribution of history to the formulation of theoretical rules. This is probably because he sees theory very much in the light of its practical function as an instructive didactic tool, and he is extremely sceptical about the instructive use of theory in the form of a fixed set of prescriptive or 'positive' rules.

A military theory should demonstrate its practical use by offering young officers reflection rather than trivial rules about the extremely complicated topic of warfare.[71] Theory should invite them to an analytic examination that leads to an understanding of the objects under investigation. There is only one way to instil this familiarity in pupils even before they engage in direct operational activities, and that is through a close study of military history.[72] History

66 See Kaye, 'Challenging Certainty', 44–5.

67 Clausewitz, *On War*, 132/Clausewitz, *Vom Kriege*, ii.1, 277; Clausewitz, *On War*/Clausewitz, *Vom Kriege*, ii.2, 291.

68 Clausewitz, *On War*, 139–41/Clausewitz, *Vom Kriege*, ii.2, 288–9; Clausewitz, *On War*, 149–50/Clausewitz, *Vom Kriege*, ii.3, 303; see also above, §4.3 and §4.5.

69 Clausewitz, *On War*, 71/Clausewitz, 'Nachricht' in Clausewitz, *Vom Kriege*, 182–3.

70 Clausewitz, *On War*, 144/Clausewitz, *Vom Kriege*, ii.2, 295.

71 Clausewitz, *On War*, 141/Clausewitz, *Vom Kriege*, ii.2, 290.

72 Clausewitz, *On War*, Clausewitz, 141, 144/*Vom Kriege*, ii.2, 290–1, 295; see also Clausewitz, *On War*, 615/Clausewitz, *Vom Kriege*, viii.8, 1007.

allows the student to relive the decisions of great generals in what Jon Tetsuro Sumida has called 'a form of psychological re-enactment'.[73] Given his didactic aims, it is not surprising that Clausewitz prefers a detailed historical account of a single campaign to more general surveys, and this in turn explains his predilection for recent and contemporary history.[74] In a letter of 1829 to Karl von der Gröben, Clausewitz explained that the instructive aim of his historical writings was the reason why he did not publish his historical works: 'I am never afraid to ask the "why" of the "why," since I do not aim to write something that is agreeable, but only to seek for myself and others unquestionable truth and instruction'.[75] History also had taught Clausewitz himself an important lesson. The crushing defeat of Prussia at the Battle of Jena in 1806 showed to him the danger of the mindless application of military procedures that had functioned well enough in the age of Frederick the Great. The antidote against such petrified 'methodism' is to develop the critical capabilities (*Kritik*) of young officers.

In criticism, history is not used for theory, but theory is used for history. The point of criticism is not that theoretical *content* is brought to bear on history, but rather that criticism is a theoretical *method* applied to history.[76] Historical criticism has three functions: it helps to clarify dubious facts, it contributes to establishing causal relations, and it inspects (*prüfen*) means in relation to ends. There is a clear relation between these functions. It is only thanks to causal criticism that we are able to isolate the topics that are worthy of evaluative criticism.[77] But the ultimate aim remains evaluative criticism in the form of 'praise and blame' (*Lob und Tadel*) about means and ends.[78] It is this evaluative function that enables criticism to be instructive (*belehrend*).[79] Given his aim of educating young officers to become future leaders, it is not strange that Clausewitz concentrates on the actions and decisions of great generals. And since he finds it important to judge individuals, he formulates a criterion of fairness that guides this judgment. The military critic should endeavour to place himself in the shoes of his historical subject, and only praise or blame him on account of what he knew or could have known.[80] The critic should try

73 Sumida, *Decoding Clausewitz*, 100; see also Engberg-Pedersen, *Empire of Chance*, 136–45. For 'reenactments' in the form of wargames see below, chapters 6 and 7.

74 Clausewitz, *On War*, 171–2/Clausewitz, *Vom Kriege*, ii.6, 338–40.

75 Quoted in Kessel, 'Zur Genesis', 421; see also Hahlweg, preface in Clausewitz, *Feldzug*, 937.

76 Clausewitz, *On War*, 156–7/Clausewitz, *Vom Kriege*, ii.5, 312–13; cf. Honig, 'Clausewitz and the Politics', 30.

77 Clausewitz, *On War*, 159/Clausewitz, *Vom Kriege*, ii.5, 317.

78 Clausewitz, *On War*, 156/Clausewitz, *Vom Kriege*, ii.5, 312.

79 Clausewitz, *On War*, 157/Clausewitz, *Vom Kriege*, ii.5, 314.

80 Clausewitz, *On War*, 164/Clausewitz, *Vom Kriege*, ii.5, 325–6.

to reproduce the mental activity of his subjects—hence Sumida's point about re-enactment. Criticism should use the same practical language as the subjects that it studies in the midst of their choices and actions. Only then will criticism exercise its instructive function and only then will this theoretical tool be adequately applied.

Moreover, if criticism is to exercise its function properly, it needs historical counterfactuals. Since criticism should have the form of an evaluative judgment about the aims and means of generals, it should not only inspect the aims that they actually formulated and the means that they actually used, but also take into consideration the aims that they *could* have formulated and the means that they *could* have used: 'Critical analysis is not just an evaluation of the ends actually employed, but of *all possible means*—which have first to be formulated, that is, invented. One can after all, not condemn a method without being able to suggest a better alternative'.[81]

A telling example is Clausewitz's analysis of Napoleon's actions in Italy after his victory against the Austrian Archduke Charles on the river Tagliamento in 1797. After this limited success Napoleon signed the Treaty of Campo Formio with the Austrians, which provided the French with rather limited gains. Clausewitz starts his critical analysis with a counterfactual: what if Napoleon had not signed the treaty, but rather had pressed on with his army to Vienna? In order to use this counterfactual for an evaluative analysis, Clausewitz considers Napoleon's knowledge about the available means to his ends. Moreover, Clausewitz discusses aims at one level that become means at the next higher level, until finally the ultimate objective of war is reached: peace. Clausewitz zooms out from the limited perspective of a local military success in Italy to an ever-wider perspective with ever more players. Had Napoleon known that the Austrians had no troops available between him and Vienna, then a continuation of the campaign might have been a good idea. However, an adequate evaluation of a continued offensive depends on the value the Austrians set on the retention of Vienna. If, rather than lose their capital, they would have preferred to grant the French more favourable terms than they actually did in the Treaty of Campo Formio, the ultimate aim of a favourable peace would have justified Napoleon's offensive. And if, in his turn, Napoleon had 'somehow' known this, then 'the critic would have no more to say'.[82] But this argument would be

81 Clausewitz, *On War*, 161/Clausewitz, *Vom Kriege*, ii.5, 321. For other examples of counterfactuals see Clausewitz, *On War*, 208–9/Clausewitz, *Vom Kriege*, iii.12, 394–5; Clausewitz, *On War*, 583, footnote/Clausewitz, *Vom Kriege*, viii.3, 958, footnote; Clausewitz, *On War*, 616/Clausewitz, *Vom Kriege*, viii.8, 1007; Clausewitz, *On War*, 627–9, 631/Clausewitz, *Vom Kriege*, viii.9, 1024–7, 1031.

82 Clausewitz, *On War*, 160/Clausewitz, *Vom Kriege*, ii.5, 319.

too simple. The critic should take an even wider point of view and ask what would have happened if the Austrians had abandoned Vienna and withdrawn their army without capitulating. In that case the French Directory might have brought in their armies on the Rhine in addition to Napoleon's Italian army. A subsequent military success of the French against the Austrians would then have been likely. What use would the Directory have made of this success? Would the French have pushed on in the hope of a complete victory? In that case they might have lost themselves in the space of the Austrian Empire. This would have led to a complete reversion of a previously favourable situation. Clausewitz has no doubt that Napoleon was aware of these wider negative ramifications. This is not the end of Clausewitz's analysis, but it is enough to show how he starts with a counterfactual, uses subsequent counterfactuals, takes into account the situational knowledge of the most important actors, and finally feels able to pass a (positive) verdict about Napoleon's decision to stop his campaign and settle for the limited gains of the Treaty of Campo Formio.[83]

Clausewitz's interest in great men and his efforts at inside knowledge are indeed typical of nineteenth-century historicism. But *On War* provides us a more specific context for his use of counterfactuals in the *Campaign of 1815*. In the latter work, counterfactuals form the vital element of a critical investigation, in analytic sections that often carry the explicit title of 'Criticism' or 'Reflection' (*Betrachtung*) and are interspersed among more descriptive sections. The point made in *On War* about the interrelated character of factual, causal, and evaluative criticism is brought into vivid practice in the *Campaign of 1815*. Moreover, in *On War* the author explains how his counterfactual criticism is an instrument in the pursuit of instructive ends; and these ends are indeed vigorously pursued in the *Campaign 1815*. Facts should be used to draw 'clear and instructive conclusions' (*ein klares und lehrreiches Resultat*).[84] Clausewitz had good reasons to use this work for teaching the Crown Prince.

5 Context: Military History in General

Although the eight functions ascribed by Nolan to counterfactuals were formulated with the aims of contemporary historians in mind, we have seen that most of these functions can be clearly distinguished in Clausewitz's *Campaign*

83 Clausewitz, *On War*, 159–61/Clausewitz, *Vom Kriege*, ii.5, 317–20; see also Gallagher, 'The Formalism', 30–1.

84 Clausewitz, *Campaign*, 73/Clausewitz, *Feldzug*, §11, 963; see also Clausewitz, *Campaign*, 192–3/Clausewitz, *Feldzug*, §53, 1089.

of 1815.[85] The instructive function of this work (as further explained in *On War*) explains why a central part is played by evaluative historical counterfactuals that assign praise and blame to individual military leaders. Other kinds of historical counterfactuals are often used as means toward the same end. In that sense, the *Campaign of 1815* serves as a rewarding case for a study of the intricate patterns formed by the interrelated functions of different types of counterfactuals. But why, actually, was it so easy to present the main stages of this account in counterfactual form? Surely this must be related to the military theme of the work. One of the first truly systematic uses of counterfactuals was indeed made by an immediate predecessor of Clausewitz, the Welsh officer and military writer Henry Lloyd (1718–1783).[86] He did not just write a *factual history* of the campaigns of Frederick the Great, but a *critical analysis* that may have been a source of methodological inspiration for Clausewitz. Catherine Gallagher notes that Lloyd's critical project entailed three levels of narrative: 'what happened, what might have happened, and what should have happened. It is in the last of these levels that Lloyd comes into the full exercise of his critical powers, and we should note that when the "ought" narrative begins, unmarked transitions take us into the realm of the counterfactuals'.[87] This perceived aptitude of military history for counterfactual treatment has remained in view ever since. In 1999 and 2001 Robert Cowley edited two popular collections of counterfactual history: *What If?* The first volume consists exclusively of military history (reflecting its original publication in *MHQ: The Quarterly Journal of Military History*). The second volume no longer focuses on military topics, but in the introduction the editor repeats a point that he already made in the first volume: 'Armed confrontation provides counterfactual history with its most natural arena'.[88]

Various reasons have been given for this aptitude of military history for the formulation of historical counterfactuals. Martin Bunzl points to the importance of aims. Counterfactuals explicitly or implicitly ascribe aims to historical actors, and counterfactuals can be used to test the rationality of the choices that were supposed to realize these aims. The advantage of military aims is that they tend to be clearer than other aims; military aims can often be expressed unambiguously in terms of *winning*. If the aims are so clear, then it becomes easier to construct counterfactuals that study their (lack of) realization. But Bunzl appreciates that this is by no means an absolute criterion; often military

85 See above §5.3.
86 See Lloyd, *War*.
87 Gallagher, 'The Formalism', 26.
88 Cowley, *What If?* (2002), xviii; Cowley, *What If?* (2001), xiii.

aims are muddled while non-military aims can be surprisingly clear.[89] A more important reason is suggested by Cowley in the introduction to his first volume of *What If?*: 'What ifs can define true turning points. They can show that small accidents or split-second decisions are as likely to have major repercussions as large ones'.[90] Military history seems especially prone to dramatic turning points, hence its special counterfactual potential.[91]

But what exactly is a 'turning point'? We have seen that a counterfactual consists of an antecedent and a consequent. In a counterfactual context, turning points can be described as antecedents that have dramatic consequents. Consider the historical counterfactual 'If Marshall Grouchy and his army corps had arrived in time, Napoleon would have won the Battle of Waterloo'. Here a French victory at Waterloo is the dramatic consequent of an antecedent turning point: Grouchy arrives in time. But is this a *plausible* historical counterfactual? The plausibility of a historical counterfactual depends on the plausibility of the antecedent turning point, which in its turn depends on the plausibility of the preceding scenario leading to the turning point itself. In the example, if it is possible to construct a plausible scenario that explains how Grouchy could have arrived in time, then we have a plausible counterfactual. Noel Hendrickson observes that counterfactual turning points should be constructed with a minimum of change in the preceding scenario that is supposed to produce the turning point in question.[92] This general 'minimal rewrite rule' can be made operational by the use of three criteria.

The first criterion for antecedent scenario selection is the amount of prior history that is perfectly preserved. This indicates how far back in time events have to be altered in order to subsequently produce the turning point in question. The shorter back in time these events, the more history is preserved.[93] The second criterion pertains to the probability of the scenario that results in the turning point. The probability of the turning point can be expressed as the product of the probabilities of each of the events in the chain given prior to it.[94] The third criterion concerns unity. The preceding scenario for a counterfactual turning point is unified to the extent that it consists of fewer starting points.[95] The upshot is that we have constructed a plausible historical

89 Bunzl, 'Counterfactual History', 852.
90 Cowley, *What If?* (2001), xii.
91 See also Kaye, 'Challenging Certainty', 52.
92 Hendrickson, 'Counterfactual reasoning', 365–86; see also Tetlock, 'Counterfactual Thought Experiments', 16–31.
93 Hendrickson, 'Counterfactual reasoning', 372–3.
94 Hendrickson, 'Counterfactual reasoning', 375.
95 Hendrickson, 'Counterfactual reasoning', 380.

counterfactual when the turning point is caused by *few* and *probable* starting points whose date is *shortly* before the turning point.

Using these three criteria for a plausible counterfactual, it is easy to see how military history can offer a myriad of counterfactual turning points that are the *immediate* and dramatic results of small contingencies or decisions. This falls under the first criterion: perfectly preserved prior history. It is also easy to see how in a military campaign, let alone in one battle, a *single* contingency or decision can have dramatic results. This falls under the third criterion: unity. So the first and the third criterion explain why military history makes such promising counterfactual potential. The second criterion on the other hand, probability, has no privileged status in military history. There is no a priori reason why counterfactuals should be more probable in this branch of history than in other fields of history.

If the three criteria for a plausible counterfactual are indeed close to the common notions of previous and contemporary historians, and if the first and third criteria form part and parcel of military history in particular, then one would expect these criteria to remain predominantly *implicit* in works of military history, while the more problematic second criterion of probability receives more *explicit* attention. This is indeed the case in Clausewitz's *Campaign of 1815*. Most of his counterfactuals fulfil the criteria of perfectly preserved prior history and unity, but the application of these criteria remains implicit. On the other hand, we have seen that much of Clausewitz's critical method in the *Campaign of 1815* is geared toward an investigation of the (im)probability of its many counterfactuals. For instance, Clausewitz does not see a probable scenario that allows a timely arrival of Grouchy at the battle of Waterloo. Hence, he thinks that 'If Grouchy had arrived in time, Napoleon would have won the Battle of Waterloo' is an implausible counterfactual.

This evaluation of the counterfactual potential of military history in general and the special case of Clausewitz's *Campaign of 1815* points to a surprisingly modern relevance of the seemingly antiquated field of military history at the operational level and its preoccupation with the motives and actions of great generals. The field seems indeed to enjoy a modest but no less surprising renaissance in the judgment of academic scholars. Robert Citino notes about operational military history: 'Once dominated by personalist modes of analysis that consisted almost exclusively of blaming General X for zigging when he should have zagged, or turning left when he should have turned right, it is now much more likely to emphasize systematic factors: the uncertainty of the battlefield (often metaphorically expressed, per Carl … von Clausewitz, as the "fog of war"), the ever-present problems of information-gathering and -sharing, and the inherently asymmetric nature of war. As historians in all fields seem increasingly willing to recognize the role of contingency, chance, and even "chaos" in historical development, operational military historians find themselves in the unusual position of being well ahead of the scholarly curve: they have been talking about all of these things for years'.[96] Clausewitz himself made significant contributions to all these 'systematic factors' (see above, chapter 4) and this explains much of his continuing relevance—although the contemporary enthusiasm generated by these topics has led to confusions that cannot be ascribed to himself.

Allan Beyerchen wrote a classic article on the role of chaos and nonlinear processes and their role as impediments to the formulation of a theory of war in *On War*, but his thesis has met with serious and pertinent criticism.[97] Terence Holmes puts the role of chaos into perspective by arguing that Clausewitz, although he appreciates the complicating influence of friction and interaction, still maintains that a military genius will be able to predict the reactions of his opponents and include these in his plans—so chaos does not make planning impossible.[98] In addition, Paul Roth and Thomas Ryckman answer Beyerchen with the more fundamental observation that chaos theory consists of concepts that 'have a specifically precise meaning only within the confines of mathematical theory', which means that 'the promised benefits of chaos theory *vis-à-vis* history are either fantastic or, at best, an extremely loose heuristic'.[99] Moreover, we have already noted how some historians tend to compound these problems, by using notions of chaos and contingency in conjunction with counterfactuals for anti-determinist claims about the nature of history. On this account as well, Clausewitz shows an enduring relevance. He does not use counterfactuals to make claims about the open or closed character of history, but rather puts them to instructive use as part of a critical method that evaluates the possibilities of means and ends in individual cases, even if this means the use of Citino's 'personalist modes of analysis' that consist of 'blaming General X for zigging when he should have zagged'.

96 Citino, 'Military Histories', 1079; see also Gallagher, 'The Formalism', 32.
97 Beyerchen, 'Clausewitz', 59–90.
98 Holmes, 'Planning', 130.
99 Roth, 'Chaos', 30–44; see also Engberg-Pedersen, *Empire of Chance*, 66.

CHAPTER 6

Models of War, 1770–1830: the Birth of Wargames and the Trade-off between Realism and Simplicity

1 Introduction

In the beginning of the history of military board games, or 'wargames', there was chess. However abstract its character, chess was originally a battle-game and had always been associated with warfare.[1] Hence it is not surprising that wargame designers first turned to chess. In the Early Modern period we see the development of games that maintained to all intents and purposes the mechanics of chess, while merely adding a few specifically military pieces to a slightly enlarged game board. These military chess variants would continue to appear between 1770 and 1830; but this period, so revolutionary in so many other regards, would also see the first *realistic* wargames—realistic in the sense that they tried to offer a model of a real military conflict. In this age of revolution, a whole range of completely new game concepts was introduced that have remained at the core of subsequent military board games and video games. In the most sophisticated early games, the abstract chess board was substituted by a huge game board that contained detailed three-dimensional topographic features upon which infantry fought melee battles, cavalry charged, and artillery delivered volleys. Units were supported by complex logistical systems, while players were kept in suspense by the fog of war and the friction of random events.

In the present chapter I discuss twelve wargames published between 1770 and 1830. I refer to each game by its author and its year of publication; for the full titles plus a brief characterization of each game see Table 4 at the end of this chapter. As they experimented within and beyond the simple format of chess, designers created more or less refined wargames, consisting of rules, units and a game board, that represented a real-world system consisting of warfare.[2] These wargames were what we now call 'models'. Since models are

1 Murray, *A History of Board Games*, 83.

2 Chess games: M.M. (1770) and Hoverbeck (1806). Wargames based on chess: Hellwig (1780); Allgaier (1796); Montbrison (1818); Champblanc (1828) Also based on chess is Firmas-Périès, *Le jeu de stratégie*. On the connection between wargames and chess see Hellwig (1780), 4, 6–12; Montbrison (1818), vi, 27; see also Perla, *The Art of Wargaming*, 15–16; Pias, *Computer*, 204; Juul, *Half-Real*, 10; Sabin, *Simulating War*, 79.

 | DOI:10.1163/9789004536678_008

less complicated than the system that they represent, they can be understood and manipulated more easily than the real system itself. If a wargame amounts to a model, then playing such a game amounts to a simulation. In this way, the term 'model' is used in a relatively passive sense and the term 'simulation' in a relatively active sense. Models are *used* to *run* a simulation, not the other way round.[3] We will see how wargame designers between 1770 and 1830 used similar concepts if not quite the same terminology.[4]

Models can be more or less realistic, and they can also be more or less simple. Ideally models are both realistic and simple, but since realism is generally bought at the price of simplicity, a good model should give, in the words of the modern author Beat Schwendimann, 'a judicious trade-off between realism and simplicity'.[5] This trade-off problem is typical of models in general and it was also very relevant for the early wargames designed between 1770 and 1830. As they ventured beyond the simple format of chess, designers quickly realized that increased realism implies increased complexity, which inevitably brings decreased playability in its wake. For wargame designers the general realism-*simplicity* problem for models tended to take the specific form of a realism-*playability* problem.[6] The balance between realism and simplicity (and hence playability) has influenced the evolution of wargames ever since, and the years between 1770 and 1830 provide us with the first solutions for this problem.

While chess has always been the subject of scholarly attention, wargames have had a more chequered career. Sometimes wargames were not taken seriously enough because of the association with amateur geeks; while sometimes they were taken too seriously, and hence classified by the military.[7] Although a full history of the wargames published between 1770 and 1830 still needs to be written, apologists of the genre have gained in strength.[8] There is now a growing appreciation of wargames as part of the history of rational thinking,

3 Schwendimann, 'What is the difference'; see also Oswalt, 'Current Applications', 156–7; Johnson, 'Simulation Modeling', 1511; Pias, *Computer*, 206; Sabin, *Simulating War*, 4; Engberg-Pedersen, *Empire of Chance*, 36; Ministry of Defence, *Wargaming Handbook*, 13: 'A model is a representation of a system, entity, phenomenon, or process. A simulation is the execution *over time* of models representing the attributes of one or more entities or processes'.

4 The concept and even the word 'model' had already been around at least since René Descartes, who used 'model' both for a material and a conceptual depiction of a more complex reality (Schuurman, 'René Descartes', 261–81).

5 Schwendimann, 'What is the difference'; see also Roher, 'Better living', 26; Köstlbauer, 'The Strange Attraction', 170; Sabin, *Simulating War*, 4–6.

6 See Engberg-Pedersen, *Empire of Chance*, 121; Sabin, *Simulating War*, 19, 27–8.

7 Sabin, *Simulating War*, 15.

8 On wargames between 1770 and 1780 see Hohrath, 'Prolegomena', 139–52; Perla, *The Art of Wargaming*, 15–34; Hilgers, *Kriegsspiele*, 43–71; Nohr, 'Strategie Spielen', 7–28; Nohr,

of their place in educational, moral, and scientific discourses, and of their role as models for strategic action both in military and civil contexts.[9]

The relevance of the realism-simplicity dilemma for wargames has been appreciated by modern authors, but to my knowledge, the present chapter is the first study devoted to solutions for this problem between 1770 and 1830.[10] I will argue that the considerable variation between the twelve games discussed in this chapter can be understood, at least partly, in terms of different solutions for the realism-simplicity dilemma.

2 Wargames, 1770–1830: Types and Functions

Eleven out of the twelve wargames from the period 1770–1830 discussed here were developed in the German-speaking parts of Europe, while number twelve is French: Montbrison (1818). I have studied these games exclusively on the basis of the books that contain the game rules. The buyers of these game manuals in most cases were supposed to produce the games themselves, or *have* them produced. Hence the manuals tended to contain detailed instructions for the construction of the games. Many games both stressed their entertaining as well as their useful function.[11] But, as might be expected, the less complicated games tended to stress fun for a bourgeois public, while the more complicated games tended to stress instruction for officers. A wargame of the former category could stress that it was nothing but a kind of 'cheerful chess' (*les échecs égayés*), while a wargame in the latter category could go so far in stressing its serious and useful character that it downplayed its character as a game at all.[12] Wargames developed between 1770 and 1830 can be divided into at least two types, which can be called symmetrical and situational, and which are connected with (partly) different functions.

Firstly, *symmetrical wargames* used symmetrical maps on which both players placed exactly the same units on exactly the same position; an example is the chess-like game of the anonymous author M.M. (1770). The realism of these

---

Johann C.L. Hellwig; Sandkühler, 'Die philantropische Versinnlichung', 69–86; Creveld, *Wargames*, 144–51; Engberg-Pedersen, *Empire of Chance* 287 n. 92.

9 Nohr, *Strategie Spielen* 10; Hershkovitz, 'Wargame business', 70–2.

10 Sabin, *Simulating War*, 98–9.

11 Military audience: Hellwig (1780), xix; Reiswitz (1812), 3; Perkuhn (1817), i (Vorrede); Aretin (1830), xxxi. Civilian audience: M.M. (1770), 8; Champblanc (1828), xiv; Montbrison (1818), i.

12 Respectively Montbrison (1818), vi and Venturini (1804), 2; see also Hohrat, 'Prolegomena', 141.

wargames was limited to the general mechanics of tactical or higher operational levels of conflict, and did not refer to any specific conflict. Symmetrical wargames could be used for recreation or for exercising young officers in the elementary mechanics of the art of war.[13] The designers of these games stressed the privileged function of wargames over military textbooks, because they were able to *show* rather than to blindly *inculcate* rules. Wargames demanded active participation rather than passive absorption. In this sense they belong to a well-entrenched Early Modern pedagogic tradition. This tradition required teachers to instruct in the simplest way possible by *showing* and students to learn by *doing*. These educational ideas had already been defended by Comenius and Locke in the seventeenth century, and counted Rousseau, Pestalozzi, and Johann Bernhard Basedow (1724–1790) among their more recent representatives, although the reception of these new ideas was far from uniform.[14] For instance, German Enlightened military thinkers tended to be more interested in education than their French counterparts. Frederick the Great stressed the idea that the military profession could be studied theoretically and hence that it required academic instruction. From the late 1770s onwards, countless regimental military schools throughout Germany appeared. The growth of these institutions stimulated the production of military literature, and this institutional context may explain the preponderance of German wargames.[15]

Secondly, *situational wargames* presented a specific military situation on a tactical or higher operational level in a specific loco-temporal framework. A start in this direction was made by Hellwig (1780), who remained faithful to a symmetric game board and evenness in the type and number of units, but who allowed for the possibility of a flexible and hence non-symmetrical setup. He was followed by Algaier (1796), who suggested that the game board itself of his small tactical wargame could be customized asymmetrically, and by Opiz (1806), who suggested the same for his large strategic game board.[16] When symmetry in the type and number of military units was also dropped, we reach the full dimensions of a situational wargame.

Situational wargames, like symmetrical wargames, could be played for recreational or instructive purposes, and these would indeed remain their dominant functions in the Napoleonic era. In the course of the nineteenth century, however, situational wargames would be used increasingly for the simulation

13 Hellwig (1780), 155; Hoverbeck (1806), 70; Opiz (1806), 42; Reiswitz (1812), 3; Perkuhn (1817), 1; Aretin (1830), xi, xxxi.

14 Sandkühler, 'Das Kriegsspiel', 69–73; Engberg-Pedersen, *Empire of Chance*, 106–7.

15 Gat, *Military Thought*, 57–68.

16 Allgaier (1796), 9; Hellwig (1780), 123, 137; Venturini (1798), 1–2; Opiz (1806), 25.

of *future* conflicts, as a preparation for these conflicts themselves. In this way, situational wargames could be used in a more direct and practical way, unlike wargames used for general instruction. In the Napoleonic era we see the development of wargames that had the *potential* to become practical (rather than merely instructive) wargames, although none of them actually fulfilled this potential. One important step in this practical direction was made by Reiswitz Sr (1812). He improved the recently introduced concept of a topographic game board (see Figure 4), and he replaced rigid (fixed) game rules with free (flexible) rules. In his game, each side consisted of a team of several officers who wrote out their orders in a separate room. These orders where then processed by an umpire (*Vertrauter*) in the central room, who moved the units on the game board in accordance with these instructions. Typically the umpire was a senior officer who was given the discretionary power to steer the game through situations that would have been too complicated for rigid rules.[17] The umpire developed the *General-Idee* of the battle and assigned specific units to the opponents in accordance with this idea. The umpire also decided which units started out of sight of the opponent, using a so-called *Truppenverberger*, and he decided when these units were revealed. In this way Reiswitz Sr (and his son in 1824) provided a degree of realism and flexibility that would be further developed in the free (rather than rigid) German *Kriegsspiel*, and that allowed the detailed formulation and testing of tactical (and later also) strategic scenarios (see the next chapter).

In addition to future conflicts, designers in the period 1770–1830 also started to experiment with wargames that could be used to simulate *historical* battles or campaigns, although this start was again rather tentative. While historical wargames are a well-represented genre in present-day board games and video games, the first historical battle that I have been able to find in the twelve game manuals is by Montbrison (1818): the Battle of Denain (1712). He depicts the real battle field and gives the unequal battle order of the French and the Dutch-Austrian armies, but he takes care to stress that 'A completely equal chance should be offered to the two parties, especially at the start of the game'.[18] Here we have the first attempt, as unconvincing as attempts in subsequent wargames, to create equal chances for victory in what is supposed to be a *historical* simulation. Actually Montbrison's game was much too simple to provide a realistic representation of a historical or future conflict, but at least Reiswitz Sr (1812) and Reiswitz Jr (1824) may have had this potential on the

17 Reiswitz Sr (1812), 2; Reiswitz Jr (1824), 26–7.

18 Montbrison (1818), 163.

FIGURE 4 The wargame of Leopold George von Reiswitz (1812)
SOURCE: STIFTUNG PREUßISCHER SCHLÖSSER UND GÄRTEN BERLIN-BRANDENBURG. PICTURE BY ROMAN MÄRZ, BERLIN

scale of separate battles, while the same could be said of Venturini (1798) and Aretin (1830) on the level of entire campaigns.[19]

3 Historical Background: Peace and War and Peace

Although it makes sense to consider the period 1770–1830 as the period when wargames were conceived and brought to a respectable level of maturity, this period, consisting of the last years of the Ancien Regime, the French Revolution, Napoleon, and the Restoration, is of course the very opposite of monolithic. It seems reasonable to suppose that these very different times must have left different marks on the twelve board games that are under investigation here, although actually it is not always so easy to assess the nature of this influence. Aretin (1830) assumes that the Napoleonic wars seriously hampered the production of wargames, but this is a doubtful assumption.[20] Of the twelve games considered here, a respectable number of five games were published between 1789 and 1815. Moreover, the onset of Prussia's clash with Napoleon in 1806 may have influenced the timing of the publication of the *National-Schach* of Hoverbeck (conceived in 1803, but published in 1806), which is

19 Aretin (1830), xix.
20 Aretin (1830), xxii.

replete with a belligerent nationalism.[21] Similarly, the coming of war in the same year may have stimulated the younger Opiz to finally publish a rather more sober wargame that his father had already developed in 1760.[22] Finally, Reiswitz Sr (1812) writes that he lost his job in the Prussian army as a direct consequence of Prussia's catastrophic defeat in 1806, but also informs us that this gave him time to develop his wargame. When he demonstrated the game to King Frederick William III, of whom Napoleon had remarked that he could talk about little else but 'military headgear, buttons, and leather knapsacks', he received the enthusiastic support of the King and his two sons, the future King Frederick William IV and the future Emperor William I.[23] The subsequent state-funded support for Reiswitz Sr (and also for Reiswitz Jr) was more than the sign of a Prussian king's love of baubles. The game was sponsored as part of a reform movement that tried to infuse military and civilian institutions with a new vigorous spirit. Reiswitz's wargame was seen as a valuable element in the programme of a state that should 'educate its way out of the danger'—and although this Prussian response was caused by the defeat of 1806, we have seen that it had firm roots in an existing Early Modern educational tradition.[24]

While the manuals for the wargames published in the extremely belligerent period between 1792 and 1815 stressed their military use, two wargames that clearly focus on civilian amusement were published outside this period: one by the anonymous author M.M. (1770) and one by the only French author, Montbrison (1818). The latter's game, published three years after the collapse of the French Empire, is presented as the French translation of a 'lost' and subsequently 'rediscovered' additional chapter of Laurence Sterne's *Tristram Shandy*, in which Tristram's father Walter presents a wargame supposedly invented by Walter's brother Toby. The collapse of the First Empire has made the French author (who had studied with Napoleon at the Military Academy in Paris) ironic about the activity of warfare. Montbrison suggests, not unreasonably, that the time has come to play and be entertained rather than to wage war.[25]

In addition to the influence of this obvious pattern of incessant warfare between 1792 and 1815, preceded and followed by more peaceful times, there is the more specific question of the influence of the French Revolution and the Napoleonic 'military revolution' on contemporary wargames.[26] Napoleon's

21 Hoverbeck (1806), see unpaginated dedicatory letter.
22 Opiz (1806), 42.
23 Creveld, *Wargames* 147.
24 See above, §6.2; see also Engberg-Pedersen, *Empire of Chance*, 114.
25 Montbrison (1818), 1.
26 See especially Béraud, *Le révolution militaire*.

preference for swift campaigns crowned with decisive battles that knocked out entire nations in a single blow has often been contrasted with military operations during the *Ancien Régime* that proceeded at a more leisurely pace and were mounted to capture (a string of) towns and strongholds. The importance attached to the conquest of individual towns (with a bonus for *not* burning them down) in the game of Allgaier (1796) seems indicative of the old style of warfare.[27] When on the other hand Reiswitz Sr (1812) presents the option of a so-called *Schlagepartie* which can only end with a decisive battle, we definitively seem to have entered the Napoleonic era.[28] Moreover, military innovation that was made possible by developments in cartography, was reflected in wargames by a transformation from the abstract chessboard to the topographic game board. On the whole, however, the specific impact of the momentous military changes in the period 1770–1830 on wargames is surprisingly light. This is probably because the nature of these changes was not only dramatic but at the same time relatively subtle. For instance, the cumbersome depot system of eighteenth-century warfare has often been contrasted with the improvised French practice of 'living off the countryside' during the wars of the Revolution. If this simply implied the difference between the presence and the absence of a depot system, then it would be relatively easy to model in wargames; but actually the difference was more complicated, especially if we contrast the Ancien Regime not with the French Revolution, but rather with the logistical system perfected by Napoleon. Although the Emperor maintained the light baggage trains of the Revolution and his armies continued to live off the countryside during the dispersed phases of his campaigns, he actually augmented and organized a depot system that allowed him to support the French projection of power deep into Central Europe with a constant flow of new ammunition and fresh recruits. While the old depot system had dictated the decisions of generals, Napoleonic 'logistical voluntarism' bent logistics to a much larger degree to the will of the commander.[29] The resulting shift in military reality was dramatic, but may have been difficult to depict in wargames. Yet although many Napoleonic innovations were not specifically incorporated in the newer wargames of our period, games at the campaign level at least allowed players to emulate such concepts as Napoleon's more dispersed advance followed by concentration for the decisive engagement during

27 Algaier (1796), 49–51; see also Hellwig (1780), 8, 122.

28 Reiswitz Sr (1812), 1.

29 Béraud, *La révolution militaire*, 63–100; cf. Creveld, *Supplying War*, 40–74; see also Clausewitz, *Vom Kriege*, v.14, 588.

a campaign; and games at the tactical level allowed Napoleonic enveloping attacks during individual battles.

4 Trade-off between Realism and Simplicity

Whatever their precise ambitions and whatever their success, it can be said that all wargame designers between 1770 and 1830 tried to create realistic models. Realism was not only claimed for complex situational games but even for simple symmetrical games.[30] The claim to realism inevitably brought the realism-simplicity problem in its wake.[31] A clear and explicit formulation of this problem was provided by Aretin (1830), who tried to develop a game 'that steers a middle course between too many rules and calculations, and too much simplicity, which would make it unlike [real] war'.[32] There was no single answer to the realism–simplicity problem and because the problem was acutely shared by most of the designers discussed here, they tended to compare their own solutions with those of their predecessors. Many manuals started with a historic overview of previous wargames. And while Early Modern polemics in general did not shy away from a rhetoric that could be exceedingly vehement by present standards, the designers of wargames tended to see their balancing act between realism and simplicity in terms of dispassionate conceptual engineering.[33]

Increased complexity in wargames, and hence decreased playability, was caused by an increase in realistic features. The realism–simplicity *problem* already showed up between 1770 and 1800 in the games of Hellwig (1780) and Venturini (1798), but the clearest *solutions* for this problem were articulated between 1800 and 1830. So, how did early wargame designers try to maintain realism while at the same time also maintaining simplicity, and hence playability for their players?

Firstly, and most obviously, designers realized that it was very important not to include unnecessary features. This implies what can be called an economic approach. Economy is stressed in particular by Montbrison (1818). The

30 Hellwig (1780), 1; Allgaier (1796), [3]; Opiz (1806), 21; Reiswitz Sr (1812), 3; Perkuhn (1817), i; Montbrison (1818), 69, 160; Reiswitz Jr (1824), ix, 15; Aretin (1830), xiii.

31 See above, §6.1.

32 Aretin (1830), xxix; see also Hellwig (1780), xv; Opiz (1806), 5–6; Chamblanc (1828), iii; Hohrath, 'Prolegomena', 147.

33 Allgaier (1796), [v–vi]; Opiz (1806), 5; Reiswitz Sr (1812), xxviii; Montbrison (1818), 237–41 n.; Champblanc (1828), iii–iv; Aretin (1830), xiii–xxix.

number of units should not be larger than *les besoins du service*.[34] When he adds rules for siege warfare, he defends this inclusion of additional detail with its indispensability.[35] He ascribes various forms of economy to 'Captain Tobie Shandy': 'Such were the true conditions of the problem. *Minimum* of time; *minimum* of space; *minimum* of units; *minimum* of difficulties. What a lot of *minimum*!'[36] In this context, Montbrison explicitly evokes Ockham's Razor, whose law of parsimony states that entities must not be multiplied beyond necessity.[37] Montbrison's favourite way of applying Ockham's Razor is the use of analogy. When he describes the ways in which pawns and sappers can take a pontoon, he feels that he does not need to introduce new rules, because earlier on he has already used the same rules for the capture of unmanned batteries.[38] Similarly, he notes the analogy between the functions of sapper, skirmisher, and light infantry, which accordingly are fulfilled by the same kind of generic unit.[39]

Once it is deemed important to bring similar cases under a single denominator, it becomes attractive to use general and abstract principles that rule multiple individual cases. Hence, many designers searched for the main principles of warfare, briefly articulated these principles, and tried to include them in their wargames. Hellwig enunciated several general principles of warfare in 1780 (which he repeated more extensively in 1803).[40] Since he assumed that he had achieved a close fit between the principles that govern his game and those that govern war in reality, he believed that it would be difficult to provide players with tips about his game without boring them with tips about the art of war itself.[41]

How convincing was this approach? If a quest for economy involving the use of analogies and general principles led to a more abstract form of realism in wargames, then the question is whether this approach was not in danger of collapsing back to the abstract level of chess from which many designers had tried to depart in the first place. Indeed, the most articulate proponent of the economic solution to the realism–simplicity problem, Montbrison, considers chess itself a true image of war.[42] And he considers his own game merely a

34 Montbrison (1818), 81–2.
35 Montbrison (1818), 133 n. 1, 202 n. 1.
36 Montbrison (1818), 37.
37 Montbrison (1818), 33, 38.
38 Montbrison (1818), 111.
39 Montbrison (1818), 58, 68.
40 Hellwig (1803), v.
41 Hellwig (1780), 144; Montbrison (1818), 26.
42 Montbrison (1818), 27.

more militarized form of chess.[43] It should be noted, however, that economy was not only propounded by designers of simple chess-like recreational wargames. Economy also figures in more elaborate games designed for a military audience. For example, Hellwig (1780) stresses at various instances the use of multifunctional units, and Perkuhn (1817) underlines the importance of analogous rules.[44] The constant dynamic tension between realism and simplicity has guaranteed the continued relevance of economy for designers of models in general, including models in the form of both simple and sophisticated wargames.

A second approach to the realism–simplicity problem was more appropriate for wargames in particular than for models in general: a change of scale from a higher operational (strategic) level to a lower tactical level. Reiswitz Sr (1812) notes that where Hellwig (1780), Venturini (1798), and Opiz (1806) had tried to enable simulations of entire campaigns, his own more limited aim is to facilitate simulations on the tactical level of separate battles.[45] He explains that his predecessors were carried away by the idea of warfare on the grand scale of entire provinces or even countries, which put players in the illustrious role of supreme commander of an entire theatre of operations. The requirement of realism on the level of an entire campaign forced designers to create mechanisms for logistics, recruitment of new units, and even peace treaties. These features increased complexity and decreased playability. Moreover, the large scale of these games meant that there was a poor match between the abstract playing units and the actual infantry, cavalry, and artillery they were supposed to model. The large time frame of strategic games had similar disadvantages.[46] Hence, according to father and son Reiswitz, *strategic* wargames combined the worst of two worlds: poor playability and limited realism. They both believed that *tactical* realism could be obtained at the cost of less complexity than *strategic* realism. They may have had a point to the extent that their solution simplified logistics, which was a notorious source of complexity. Logistical aspects are more important on the strategic level that models an entire military campaign than the tactical level that models a single battle. Hence it is not surprising that the four strategic games considered here, those by Hellwig (1780), Venturini (1798), Opiz (1806) and Aretin (1830), have

43 Montbrison (1818), ii.

44 Hellwig (1780), xi, 104; Perkuhn (1817), 29, 31; see also Engberg-Pedersen, *Empire of Chance*, 120.

45 Reiswitz Sr (1812), xiv.

46 Reiswitz Sr (1812), xxiii; Reiswitz Jr (1824), viii.

sophisticated rules for logistics, while none of the remaining games, which all model a tactical rather than a strategic level, have a logistic mechanism.

Typically, the Reiswitz switch from the strategic to the tactical level did not remain unchallenged. When Aretin (1830) defended his strategic wargame, he remarked that the Reiswitz solution resulted in a tactical game that was actually too complicated and tedious, while its limited scale meant that it missed the very essence of warfare, which for Aretin consists in (strategic) operations on a Napoleonic scale rather than (tactical) battles—hence the very title of his game: *Strategonon*.[47] Indeed, as a solution to the realism-simplicity problem, zooming in from a higher to a lower level seems of limited value. Zooming in can free the designer of some *higher-level* complexities, but if he wants to maintain a similar level of realism, we would expect him to include additional *lower-level* features. And if each level has its own complexities, then the result of zooming in from a higher to a lower level will not necessarily result in an overall reduction of complexity. However, father and son Reiswitz simply did not include as many new lower level complexities as they shed higher lever complexities. In this way they could indeed end up with less complexity and greater playability.

A third and more convincing strategy to solve the realism-simplicity dilemma was again developed by father and son Reiswitz. I already remarked on the great flexibility of their sophisticated games, thanks to the role of the umpire, who was able to solve many problems by *ad hoc* decisions, so that no detailed game rules were required.[48] Consequently, compared to games with a similar level of sophistication, the rules of the Reiswitz games are remarkably short and sketchy—not because complexity was reduced, but because it was transferred from the players to the umpire.[49] While the rulebook of his game remained slender, Reiswitz Jr demanded from its players real military knowledge 'of the use of different arms when these are put to a single purpose in an interconnected way'.[50] This military knowledge was to guide their actions in the game and should help them to avoid silly and unlikely moves. Reiswitz Jr remarks that such forms of bad play could have been limited by proliferating the rules, but this would have removed the game from the main issues (*von den Hauptsachen*).[51] This is where the umpire stepped in. Father and son Reiswitz

47 Aretin (1830), xxvi; see also Hohrath, 'Prolegomena', 148.

48 See above, §6.2.

49 See also Reiswitz Jr (1824), ix. I have not been able to consult the tables that were mentioned in Reiswitz Jr (1824) in the 'Buch-Binder Nachrichten'; see also Anonymous, 'Supplement zu den bisherigen Kriegsspiel-Regeln', 68–105; Pias, *Computer*, 220–4.

50 Reiswitz Jr (1824), 37.

51 Reiswitz Jr (1824), 37.

did not need complicated rules to prevent unrealistic moves; such moves could simply be ruled out by the umpire. The succinctness of the rules was compensated for by the substantial discretionary and improvisational powers of the umpire. The umpire acted as a 'rule sink'. In this way an original contribution was made towards the solution of the realism-simplicity dilemma. *For the players* the umpire very effectively decreased the level of detail and hence the level of complexity.[52] At the same time the umpire actively maintained a level of realism, including an explicit ban on unrealistic moves.[53]

5 The Realism-Simplicity Trade-off and Chance

Warfare is complicated by the fact that it consists of elements that are difficult or even impossible to control and for which we have seen Clausewitz using the term 'friction'.[54] Friction tends to consist of uncertainties, and some designers saw the simulation of uncertainty as an important contribution towards the realism of their games, even if this resulted in a more voluminous rule book. In this sense uncertainty is very relevant for the realism-simplicity problem. An important way to simulate uncertainty was the use of dice. Dice as generators of random events in wargames were first introduced by Opiz (1806). He observed that warfare cannot be imitated without the inclusion of chance, hence his introduction of 'The Yes and No of the dice'.[55] He used this feature right away for a variety of different concepts. For instance, dice were used to determine the position of the game board (optional) and the chances of success when players tried to move units into rough terrain.[56] For units that were out of supply, a die was first rolled to check if they would desert and, if this was

---

52 Reiswitz Jr (1824), 4; see also Pias, *Computer*, 208 n. 48 and Sabin, *Simulating War*, 31.

53 The complexity absorbing role of Reiswitz's umpire in many ways anticipated a similar feature of (military) video games. In this context Philip Sabin rightly remarks that computer games seem to minimise the trade-off between (detailed) realism and playability (Sabin, *Simulating War* 19). And when James Dunnigan, *Wargames Handbook*, 73, presents a list of differences between computer wargames and their paper predecessors, he mentions easier rules as a distinctive characteristic of computer (war)games, because the rules of computer games have been embedded in the programme itself. In addition Dunnigan stresses the ability of computer games to free the hands of the players by acting as a super clerk. In the Reiswitz games, rules were embedded in the umpire, who also acted as a super clerk; see Pias, *Computer*, 208 n. 48; Köstlbauer, 'The Strange Attraction', 176; Perla, *The Art of Wargaming*, 42.

54 See also above, §4.3.

55 Opiz (1806), 6.

56 Opiz (1806), 46–7, 61–2; see also Engberg-Pedersen, *Empire of Chance*, 128–33.

indeed the case, a second roll determined the level of losses by desertion.[57] A similar but more sophisticated serialisation of randomisation was used to determine battle results. Opiz used a combat results table that anticipates the complex professional German wargames of the 1870s and also the sophisticated recreational military board games of the 1980s.[58] Randomisation with dice was also used by Reiswitz Jr (1824) but was largely limited to the assessment of battle results, for which he used a combat results table as well.[59]

Intriguingly, the use of dice in wargames in 1770–1830 was never generally accepted, and remained the topic of a lively discussion in the relevant literature. This discussion about chance actually started already before the publication of Opiz's innovation in 1806. In 1803 an anonymous author, 'Kr'., published a review of the wargame by Venturini (1798) in which he regrets the absence of chance which is such a vital element of war.[60] Venturini had indeed stressed that a good wargame only depends on the will of its players, and is not determined by random factors; there should be no place for differences caused by 'the bravery of the troops, or betrayal, or the random favour of the weather'.[61] Similarly, when Hellwig in the 1803 reprint of his game discusses the dual aim of his wargame—instruction and recreation—he expresses an aversion to forms of chance that are not caused by the moves of the players themselves.[62] Finally, when Aretin (1830) discusses Opiz's introduction of chance, he regrets that in this game even the smartest game manoeuvres can be utterly ruined by the mere throw of the die.[63] Aretin was quite able, however, to separate his wish to reduce gamer frustration from harsh reality itself. He fully acknowledges the importance of chance in the reality of warfare, which he supports with a Napoleonic quotation about the ability to profit from chance as a defining trait of the good commander. Yet when he explicitly tries to maintain an equilibrium between realism and simplicity he argues that this middle course should exclude simplicity in the form of chess-like moves, but also complexities in the form of, amongst other things, the use of dice.[64] Here again, as in so many other cases, the trade-off between realism and simplicity in wargames between 1770 and 1830 was dynamic and unstable. It was the object of constant

57 Opiz (1806), 79.
58 Opiz (1806), 74–6, 15–16; see also Reiswitz Jr (1824), 34, 52; and below, §7.4 Figure 5.
59 Reiswitz Jr (1824), 34, 52.
60 Kr., review of Venturini (1798), 550.
61 Venturini (1798), 1.
62 Hellwig (1803), iii.
63 Aretin (1830), xxi.
64 Aretin (1830), xxix, 134; see also Hellwig (1803), v.

polemics and fruitful experimentation, as indeed it has remained down to the present day. Given this context it is not surprising that between 1770 and 1830 the fate of chance itself remained highly erratic.

Moreover, in addition to dice there were other ways to simulate uncertainty. One kind of military uncertainty concerns the whereabouts of the enemy. This 'fog of war' could be simulated during set-up of the game with a screen, or by a device that covered units when play had started. Set-up screens were introduced by Allgaier (1796); we have seen that Reiswitz Sr (1812) introduced a *Truppenverberger*; while Perkuhn (1817) used a similar mechanism.[65] Actually, playing wargames has inherent uncertainties, even without any deliberate attempt at simulation by the designers. These uncertainties are produced by interaction between the various mechanisms of the game and between the players themselves. Perkuhn (1817) notes that each type of unit in his game is unique in either its movement, or in its combat strength (*Übermacht*), or in both combined, while the interactions between these elements produce uncertainties that are reinforced by different types of terrain.[66] M.M. (1770) similarly notes the uncertainty produced by the mere *caprice des joueurs*.[67] Hoverbeck (1806) asserts that his chess like and dice-less game will stimulate young officers to think about the intentions of their opponents. He assumes that this way of thinking about '*possibility—and probability*' will foster '*composure* or *presence of mind*' (italics by the author).[68] Aretin (1830) also mentions the important lessons of cool-headedness that can be learnt from his dice-less game, especially after an *unexpected* loss.[69]

These cases of dice-less uncertainty in wargames may be typical of what Jesper Juul, a modern author on video games, calls games of emergence. While manuals for *games of progression* concentrate on the order of actions in what typically constitutes a story line, manuals for *games of emergence* will concentrate on rules of thumb that help the player choose between options that will rapidly amount to a large game tree. Games of emergence, for instance chess and other conflict games, are characterized by a relatively small number of rules that yield a relatively large number of options that can only be dealt with by deep strategies, resulting in the paradox 'easy to learn but difficult to master'.[70] And indeed Philip Sabin observes that the mere presence of dynamic

65 Allgaier (1796), 52–3; Reiswitz Sr (1812), 32; Perkuhn (1817), 37–8; see also Sabin, *Simulating War*, 106–15; and above, §6.2.

66 Perkuhn (1817), iii (Vorrede); see also Aretin (1830), 61–2.

67 M.M. (1770), 58.

68 Hoverbeck (1806), 154.

69 Aretin (1830), xxxii, 61–2.

70 Juul, *Half-Real*, 73, 57, 71.

interaction between players in otherwise simple wargames generates infinite amounts of uncertainty, even without the use of any dice.[71] And if 'difficult to learn' is already a characteristic of simple strategy games, than this certainly holds true for more complex strategy games. This modern discussion may help us to understand why so many Napoleonic designers of 'games of emergence' were able to see uncertainty and surprise in games that did not use mechanisms for the explicit generation of these phenomena. This may explain why Opiz's creative introduction of dice in his wargame was not universally imitated, and why the polemics about the simulation of uncertainty in general and chance in particular continued.

6 Discussion: Empire of Chance?

The first wargames considered in this chapter were developed during the 'High' Enlightenment while the last were produced during the high days of German Romanticism and Idealism. The Enlightenment seems to have left a clear mark on the work of even the last of the twelve game designers discussed in this chapter. Military Enlightenment thought included a belief in the desirability and possibility of the formulation of fixed and universal principles of warfare. There was also a widely shared belief in the predominantly mechanical character of these principles: warfare tended to be formulated in physical terms of space, movement, and force. Immaterial factors such as the genius of the commander and the importance of chance were appreciated as well, but were not deemed susceptible to scientific treatment.[72] These views did not end with the wars of the French Revolution and Napoleon; on the contrary, they received their most eloquent expression by Antoine de Jomini (1779–1869). He served Napoleon as a general until he defected to the Russians in 1813, but continued to regard the Emperor's campaigns as the very epitome of the universal principles of war for the rest of his long life.[73] Jomini's main principles can easily be expressed in mechanical terms and revolve around the art of organizing, moving, and concentrating military units as swiftly as possible.[74] Thus it can be said that Enlightened military thought survived well into the nineteenth century. Moreover, this way of thinking fitted hand-in-glove with a design strategy that reduced complexity by focussing on game features that expressed

71 Sabin, *Simulating War*, 55–7, 110–11, 118.

72 Gat, *Military Thought*, 130.

73 Gat, *Military Thought*, 73, 84, 115, 121.

74 Cf. Engberg-Pedersen, *Empire of Chance*, 160.

essential principles rather than trivial details, especially if these principles were of a predominantly mechanical nature.[75] Even the last designer in our period, Aretin (1830), was a firm believer in the existence of universal and fixed military principles.[76] When he argues for this point he explicitly quotes Jomini and other Enlightened sources.[77]

The influence of *post*-Enlightened thought on wargames published between 1770 and 1830 is more difficult to assess, if only because the most trenchant criticism of Enlightened military thought, Clausewitz's unfinished *Vom Kriege*, was not published (posthumously) until 1832, i.e. after the period under consideration.[78] Where Jomini tended to concentrate on operational principles and their validity throughout history, Clausewitz stressed the importance of differences caused by different historical contexts. Moreover, he was not so much interested in the formulation of prescriptive principles, as in a more philosophical analysis of interaction, escalation, friction, the relationship between means and ends and the relationship between politics and war.[79] Although he reacted against Jomini, his earliest victim was Adam Heinrich Dietrich von Bülow (1757–1807), whose work he attacked forcefully in a review that was published in 1805. Bülow in many ways represented eighteenth-century military thought, although his use of geometry and his insistence on the importance of the precise form of the triangle formed by an army's point of attack and the line formed by its base, was rather more idiosyncratic.[80] In his early review of Bülow, Clausewitz formulated his aversion to pedantic imitations of mathematics and stressed the importance of moral (psychological) forces. With this attack Clausewitz contributed to a subsequent stereotype of Enlightened military thought down to the present day.[81]

Moreover, we have already seen how Clausewitz defended the use of counterfactuals in historical case-studies as a form of vicarious learning in the instruction of young officers.[82] Another form of vicarious instruction can be obtained by playing wargames. Here again young officers can learn by stepping in the shoes of great commanders. Actually, a historical wargame can be used to test the plausibility of an historical counterfactual like 'If Marshall Grouchy

75 See also above, §6.4.
76 Aretin (1830), 133.
77 Aretin (1830), 31.
78 See Paret, *The Cognitive Challenge*, 121; Gat, *Military Thought* 125–127.
79 See above, chapter 4; see also Paret, *The Cognitive Challenge*, 121; Gat, *Military Thought*, 125–7.
80 Gat, *Military Thought*, 81–96, 131.
81 Gat, *Military Thought*, 95.
82 See above, §5.4.

and his army corps had arrived in time, Napoleon would have won the Battle of Waterloo'. This link between historical wargames and historical counterfactuals has been noted before. For instance, Philip Sabin woefully notes that 'traditionalist scholars in the humanities may see wargames as combining the worst features of game theory and counterfactual history'.[83] Whether 'traditionalist scholars' are right about the negative implication of this comparison is dubious, but at least they are correct about the connection itself. Given this connection, and given Clausewitz's eager use of historical counterfactuals for vicarious learning, we might expect him to be equally positive about the use of professional wargames as tools toward the same aim. He taught strategy to the sons of King Friedrich Wilhelm IV and we have seen that these princes very much liked the wargame of Reiswitz Sr. Yet Clausewitz himself wrote next to nothing about games, and what he did write was actually highly critical. Interestingly, he makes two comparisons with games in the context of his attack on Bülow. By leaving out moral forces that can decide the fate of battles or entire campaigns, Bülow changed war in 'a mere card game' (*ein bloßes Kartenspiel*).[84] In an early draft for *On War*, on the other hand, Clausewitz compares war not with a card game but with a game of chess.[85] Again the context is an attack against Bülow's belief that warfare can be reduced to a simple calculation which aims to concentrate a maximum of force on a single point. Since Clausewitz associates these games with antiquated Enlightened thought, it is not surprising that he finds little of use there for his own Romantic theory of war.

For a fascinating contemporary study on the intellectual paradigm shift between Enlightened and Romantic military thinking we can turn to *Empire of Chance: The Napoleonic Wars and the Disorder of Things* (2015) by Engberg-Pedersen. The author of this study starts with a well-known depiction of eighteenth-century military theory guided by geometry and mathematical calculations for the construction of star-shaped fortifications and the drilling of troops. He then observes how around 1800 the vast expansion of military operations amounted to what the Prussian officer Georg Heinrich von Berenhorst (1733–1814) called an *empire du hasard*, a hazardous empire of chance, replete with errors and uncertainties.[86] Engberg-Pedersen analyses the intricate but dramatic interactions between military events and the intellectual climate. He notes that 'Epistemology suffered a concussion at Austerlitz, at Wagram, at

83 Sabin, *Simulating War*, 16.
84 Clausewitz, *Bemerkungen*, 79.
85 Clausewitz, *Niederschriften des Werkes 'Vom Kriege'*, 655–9.
86 Engberg-Pedersen, *Empire of Chance*, 3–4, 39.

Borodino, and the effects of the impact can be measured in the texts that subsequently tried to make sense of the situation'.[87] He also notes that the epistemological 'concussion' caused by the new way of war made experience more artificial, in the sense that 'second-order' phenomena such as maps, games, and texts were increasingly used as instruments 'to replicate the contingencies of war while excluding its dangers'.[88] In this context the author pays due attention to the introduction of 'the Yes and No of the dice' by Opiz (1806).[89]

The degree to which the wargames studied here were actually affected by the contingent dimensions of Engberg-Pedersen's epistemological 'concussion', however, remained curiously restricted. We have already noted that Opiz's son may have responded to the needs of a changed intellectual climate when he finally published the 'random wargame' developed by his father in 1760.[90] But when Opiz Sr introduced chance in 1760 and played his game with eight young fellow Jesuits in Klattau (Bohemia), huddled up in a room which they called their *Museum*, it is difficult to imagine him at the epicentre of an intellectual upheaval.[91] On the other hand, we have seen that the dramatic year 1806 also saw the publication of a game by Hoverbeck.[92] It is easy to associate its nationalism with the dramatic events of the times; nevertheless Hoverbeck's pseudo-chess game does not have the faintest trace of a concussion-induced random mechanism.

The difference between the game developed by Reiswitz Sr in 1812 and the version developed by Reiswitz Jr in 1824 is instructive in a similar way. Dice were only introduced by the younger Reiswitz, nine years after the Napoleonic wars. It is possible of course that the concept was already developed by his father, but in the introduction to his 1812 game Reiswitz Sr duly and perceptively acknowledges the novel random trait in Opiz's game, without showing any readiness to include dice in his own game.[93] So contrary to what might be expected, none of the games considered here provides a clear link between the

87 Engberg-Pedersen, *Empire of Chance*, 5. This impact formed the context for Hegel's views on the development of the world spirit through war. Hegel conceived warfare as a realm of absolute chance, and this conception of chance gave him a metaphorical framework for the contingency of empirical phenomena in general, which in its turn helped him to make an absolute difference between the realms of contingency and necessity. See Engberg-Pedersen, *Empire of Chance*, 54.

88 Engberg-Pedersen, *Empire of Chance*, 115.

89 Engberg-Pedersen, *Empire of Chance*, 128.

90 See above, §6.3.

91 Opiz (1806), 41.

92 See above, §6.3.

93 Reiswitz Sr (1812), xxiii–xxiv.

introduction of dice and the turbulent military and intellectual climate of the Napoleonic wars. The realism-simplicity problem stimulated a preoccupation with clear principles. This preoccupation prolonged Enlightened thought in wargames well past its expiration date in other sectors.

So the link between a Napoleonic epistemological concussion and the introduction of dice in the wargames discussed here seems rather spurious. Interestingly, there is a much clearer connection between the (related) phenomenon of contemporary developments in statistics and probability calculation on the one hand, and the latest wargames discussed here on the other hand. In the seventeenth and eighteenth century, thanks to the increased quantity and quality of demographic data, the discipline of 'political arithmetic' started to assess individual and seemingly random events in the context of their average rate of occurrence. La Place and Poisson showed how randomness on the individual level was smoothed out on the collective level of large numbers.[94] Chance could be tamed by the laws of probability. The result was something between complete randomness and the highest degree of scientific certainty achieved by, for instance, Newtonian physics. Engberg-Perdersen very perceptively stresses the importance of this 'middle realm of knowledge, a tremendously complex epistemic field of probabilities, possibilities, conjectures, averages, modalities' for the age of the Napoleonic wars.[95] So, in the same period that chance and randomness seemed to make its most dramatic appearance on the stage of history, the Poisson curve and probability calculus showed how it could be controlled. And although an author like Clausewitz certainly stressed the role of chance in war, he also stressed the need to calculate probabilities in war.[96]

The middle realm of probability calculation is especially well represented in the late games of Reiswitz (1824) and Aretin (1830). Very typically, here we see again how a similar context allowed for both the absence and the presence of dice. We have already seen that Reiswitz Jr (1824) used dice to assess combat battle results.[97] He used statistical data about day's marches and fire power of infantry, and the range of artillery of different calibre to compose combat

94 Ball, *Critical Mass*, 53–95.

95 Engberg-Pedersen, *Empire of Chance*, 4.

96 In this way Clausewitz's ambitions to control the vagaries of war actually exceeded those of his Enlightened predecessors. Whereas chance events were preferably left out of the matrix of eighteenth-century military theory, Clausewitz believed that these events, including psychological events, could be made subject to a *Wahrscheinlichkeitskalkül*. See Clausewitz, *On War*, 85/Clausewitz, *Vom Kriege*, i.1.19, 207; Clausewitz, *Bemerkungen*, 81; but cf. Sandkühler, 'Das Kriegsspiel', 37.

97 See above, §6.4.

results tables, in which the result was determined by the throw of one or more dice. Since in reality there is statistical variation around certain mean combat results, he tried to pay tribute to the requirement of realism by simulating this variation with the use of dice, but in strict accordance with the laws of probability.[98] To this purpose Reiswitz developed calibrated combat results tables.[99]

An even more impressive use of contemporary sources for the extraction of reliable data was made by Aretin (1830), who stresses that such knowledge is indispensable for modern officers.[100] At the same time Aretin is not very interested in the simulation of statistical variation around mean values. He just takes the mean values. If sources have taught him that a certain type of canon firing at a certain distance, with a certain calibre, at a group of soldiers lined up in a certain density, has a certain average kill ratio, then he uses this mean value to calculate the battle outcome directly, without dice, and hence without statistical variation around the mean value. He is aware that this is a simplification compared to the use of dice, but probably he thinks that he can be excused for this greater simplicity, because it increases the playability of his game. Moreover, he seems to suggest that on the level of his strategic game the statistical variation in the outcome of separate tactical clashes is not sufficiently relevant to merit a dice mechanism.[101]

7 Conclusion

Designers of wargames between 1770 and 1830 had a clear model claim, i.e. in most cases they tried to develop a game that represented a real military system. They preferred models that were not only realistic, but also simple, since simplicity was a major condition for playability. In this way they quickly learned that good models demand a trade-off between realism and simplicity. They developed several solutions to this realism–simplicity problem, thus influencing the subsequent modelling of military conflict. The design of wargames, like the development of all models, was subject to both external and internal dynamics. By external dynamics I mean the dynamics of the 'real' system that the model tries to represent. Internal dynamics, on the other hand, include the influence of the realism–simplicity problem in the design process itself. A dramatic and highly relevant kind of external dynamics was provided

98 Reiswitz Jr (1824), 8–10.
99 See also Pias, *Computer*, 221; Perla, *The Art of Wargaming*, 25.
100 Aretin (1830), xxxi, 23, 62, 38, 100.
101 Aretin (1830), xxix, xxxiii.

by the Napoleonic wars. I have concluded, however, that the *direct* influence of the Napoleonic military revolution on the design of wargames between 1770 and 1830, although discernible, was relatively modest. Concentrating on the topic of chance, I have made a similar point about the *indirect* influence of the Napoleonic wars through the medium of the intellectual climate of the period on the design of wargames. Against this modest direct and indirect influence of *external* dynamics, I have noted a robust influence of design-*internal* dynamics in the form of the realism–simplicity problem. This relative predominance of design-internal aspects over external dynamics might cast some doubt on the wider historical relevance of the topic—but this was an age of fruitful experimentation. In the next chapter we will see how a direct offshoot of this period of trial and error, the free *Kriegsspiel*, became an integral part of military operational planning in Germany (and other countries), thus ensuring a substantial historical relevance of the seemingly innocuous topic of wargames between 1770 and 1830.

TABLE 4 Wargames, 1770–1830

M.M., *Le Jeu de la guerre, ou raffinement du jeu des echecs/Neues Kriegsspiel, oder: verbessertes Schachspiel* (Prague: F.A. Hoechenberger, 1770) = **M.M. (1770)**	Elementary variation on chess. Aim of the game: amusement for a general audience. Designer unknown.
Johann Christian Ludwig Hellwig, *Versuch eines aufs Schachspiel gebaueten taktischen Spiels von zwey und mehreren Personen zu Spielen* (Leipzig: S.L. Crusius, 1780) = **Hellwig (1780)**; reprint: Hellwig, *Das Kriegsspiel: ein Versuch die Wahrheit verschiedener Regeln der Kriegskunst in einem unterhaltenden Spiele anschaulich zu machen* (Braunschweig: K. Reichard, 1803) = **Hellwig (1803)**	Elaborate strategic game on a game board consisting of 1,617 squares that express topographic characteristics; 8 types of units; non-combat assets include bridges and trenches; chess-like mechanism for movement (only one unit or group of unit moves per turn) and combat; sophisticated logistics in the advanced game version, with supply lines traced to depots. Aim of the game: instruction and amusement for a military target group. Hellwig (1743–1831) was a mathematician and entomologist who taught at the military academy of Braunschweig and the university of Helmstedt.

TABLE 4 Wargames, 1770–1830 (*cont.*)

Johann Algaier, *Der Anweisung zum Schachspiel zweyter Theil* (Wien: F. J. Rötzel, 1796) = **Allgaier (1796)**	Variation on chess on an expanded game board of 600 topological squares. Aim of the game: amusement for a general audience. Algaier (1763–1823) was a German-Austrian chess master and theoretician.
Georg Venturini, *Beschreibung und Regeln eines neuen Krieges-Spiels* (Schleswig: J.G. Röhtz, 1798) = **Venturini (1798)**; reprint: Venturini, *Darstellung eines neuen Kriegsspiels* (Leipzig: Johann Conrad Hinrichs, 1804) = **Venturini (1804)**	Elaborate tactical game on a game board with 2,160–6,912 topographic squares; wide array of different types of units; seasonal factors influence movement and combat; sophisticated logistics, with supply lines traced from capital plus corn harvested for additional supplies. Aim of the game: instruction for a military target group. Georg Venturini (1772–1802) was an officer in the army of the Duke of Braunschweig and a military writer.
C.E.B. von Hoverbeck, *Das preussische National-Schach* (Breslau: Stadt-Buchdruckerey, 1806) = **Hoverbeck (1806)**	Elementary chess-like game on 121 identical squares, with the standard pieces of chess plus batteries and fusiliers. Aim of the game: instruction for military target group. Hoverbeck was a Prussian cavalry captain.
Johann Ferdinand Opiz, *Das Opiz'sche Kriegsspiel, ein Beitrag zur Bildung künftiger und zur Unterhaltung selbst der erfahrensten Taktiker.* (Halle: Hendels Verlag, 1806) = **Opiz (1806)**	Elaborate strategic game with several modular game boards with 2,304 topographic squares and 32 types of units; extensive use of dice randomizes aspects of unit movement, battle results, and logistics. Aim of the game: instruction and amusement for a military and civilian target group. Johann Ferdinand Opiz (1741–1812) was an Austrian civil servant and writer; his game was published by his son, the painter Johann Ferdinand Opiz (1775–1841).

TABLE 4 Wargames, 1770–1830 (*cont.*)

Georg Leopold von Reiswitz, *Taktisches Kriegsspiels oder Anleitung zu einer mechanischen Vorrichtung um taktische Manoeuvres sinnlich darzustellen* (Berlin: Gebrüdern Güdicke, 1812) =**Reiswitz Sr (1812)**	Sophisticated tactical game with a customizable topographic game board and countless military and support units; players produce written orders which are processed by an umpire; each move simulates a fixed amount of time (1 minute); movement measured with special compass; flexible victory conditions. Aim of the game: instruction for a military target group. Reiswitz Sr (1764–1828) was a Prussian officer and civil servant.
Johann Gottlieb Perkuhn, *Beschreibung eines Kriegsspiels zum Gebrauch für Militairs* (Hamburg: Appel, 1817) = **Perkuhn (1817)**	Compact but sophisticated tactical game on a customizable game board consisting of 1,024 cubes, each of the six sides of each cube expressing a different topographic feature (plain, hill, mountain, forest, water, town); 9 types of units; generals act as sources of communication; prisoners can be exchanged. Aim of the game: instruction and amusement for a military target group.
[Louis Simon Joseph Bernard de Montbrison], *Le Jeu de la guerre de terre et de mer, et les derniers chapitres de Tristram Shandy, trouvés dans les papiers d'Yorick* (Paris: Goujon, 1818) = **Montbrison (1818)**	Tactical chess-like game on a game board with 99 topographic squares; 16 types of units; noncombat assets include treasury and bridges; the game allows simple simulation of historical battles. Aim of the game: amusement for a general audience. Montbrison (1768–1841) first was a French officer, and in 1810 Napoleon appointed him Chancellor of the University of Strasburg.
[Georg Heinrich Rudolf Johann] von Reisswitz [*sic*], *Anleitung zur Darstellung militairischer Manöver mit dem Apparat des Kriegs-Spiels* (Berlin: [Trowitzsch], 1824) = **Reiswitz Jr (1824)**	Improved version of Reiswitz Sr (1812); the scale of the squares of the game board is smaller; each move simulates 2 minutes; and battle results are influenced by dice. Aim of the game: instruction for a military target group. Reiswitz Jr (1794–1827), son of Reiswitz Sr, was a lieutenant in the Prussian army.

TABLE 4 Wargames, 1770–1830 (*cont.*)

Franz Dominik Champblanc, *Das Kriegsspiel, oder das Schachspiel im Großen* (Wien: H.F. Müller, 1828) = **Champblanc (1828)**	Tactical chess-like game with customizable game board consisting of 460 topographic cubes; 11 types of units; each turn consists of 2 moves. Aim of the game: amusement for a general audience.
Wilhelm Freiherr von Aretin, *Strategonon, Versuch die Kriegsführung durch ein Spiel anschaulich darzustellen* (Ansbach: Dollfuss, 1830) = **Aretin (1830)**	Sophisticated strategic game with a customisable game board consisting of 144 topographic cubes; 8 types of units; generals act as sources of communication. Aim of the game: instruction and amusement for a military target group.

CHAPTER 7

Preparing for War: Prussian–German Professional Wargames and the Leadership Concept of Mission Tactics, 1870–1880

1 Introduction

Professional wargames (*Kriegsspiele*) received a major boost after the Prussian successes during the German Wars of Unification (1864–1870) and were subsequently introduced by the armies of other European powers, the United States, and Japan. They continued to play a vital role in the great campaigns of the twentieth century.[1] I provide a descriptive analysis of the main forms of Prussian–German wargames during the key decade between 1870 and 1880.[2] I then argue that the success of German wargames can be understood in the context of the military concept of mission tactics (*Auftragstaktik*). I will show how both wargames and mission tactics were driven in their turn by the even wider contexts of technological revolution in the fields of firearms and railway transport. I will argue that these contexts ushered forth professional wargames along an initially tenuous trajectory, before they became a key instrument in

---

1 Young, *A Survey*, 23–91.

2 Edmund Edler von Mayer, *Eine Studie über das Kriegsspiel* (Wien: Verlag des militärwissenschaftlichen Vereines, 1874); Klemens Wilhelm Jacob Meckel, *Studien über das Kriegsspiel* (Berlin: E.S. Mittler, 1873); Klemens Wilhelm Jacob Meckel, *Der verbesserte Kriegsspiel-Apparat* (Berlin: Vossische Buchhandlung, 1875); Klemens Wilhelm Jacob Meckel, *Anleitung zum Kriegsspiele. Erster Theil: Direktiven für das Kriegsspiel* (Berlin: Vossische Buchhandlung, 1875); Naumann, *Regiments-Kriegsspiel: Versuch einer neuen Methode des Detachments-Kriegsspiels* (Berlin: E.S. Mittler, 1877); Neumann, *Directiven für das Festungs-Kriegsspiel* (Berlin: Vossische Buchhandlung, 1872); Lebrecht Ernst Michael Thilo von Trotha, *Anleitung zum Gebrauch des Kriegsspiel-Apparates zur Darstellung von Gefechtsbildern mit Berücksichtigung der Wirkung der jetzt gebräuchlichen Waffen* (Berlin: E. S. Mittler, 1870); W. von Tschischwitz, *Anleitung zum Kriegs-Spiel* (fourth edition, Neisse: J. Graveur, 1874); Julius Verdy du Vernois, *Beitrag zum Kriegsspiel. Mit einem Plane* (Berlin: E.S. Mittler, 1876); Carl Zipser, *Anleitung zur Darstellung militärischer Manöver mit Hilfe des Kriegsspiel-Apparates, bei Zugrundelegung der neuesten taktischen Normen* (second edition, Josefstadt: C. Hoser Trautenau, 1876). On German professional wargames in the 1870s see Hilgers, *Kriegsspiele*, 92–6; Pias, *Computer*, 41–5; Creveld, *Wargames*, 151–8.

 DOI:10.1163/9789004536678_009

training and planning for war in the hands of the Great General Staff of the Prussian and hence the German army.[3]

2 Prussian–German Wargames, 1870–1880

In the previous chapter we have seen that the best known early German professional wargame was designed by Georg Leopold von Reiswitz in 1812, that he used a topographical map board, that he introduced umpires who led the game, and that the game was updated by his son Georg Heinrich Rudolf in 1824, who introduced dice to simulate chance and uncertainty in combat resolution.

The wargames used by the Prussian army in the decades following the Napoleonic Wars (until around the 1870s) continued to follow the design of the Reiswitz games. Most wargames in this period were designed by Prussian officers with the aim of training colleagues of different ages and different levels of seniority.[4] Professional wargames were also increasingly used to test specific military plans. While the Reiswitz games had been played on a small tactical scale that was deemed fit for junior officers, at a later stage wargames were staged on larger scales, including the operational level of entire campaigns. These operational games were played by officers of the Great General Staff and senior officers of the field army.[5] Operational games also included conceptual mechanisms that modelled provisions, munitions, and reinforcements.[6] The target group of serious wargames in the Napoleonic era still included recreational players, but the designers of professional wargames used by the Prussian army took care to emphasize the seriousness of their games. Their games should not be played for the sake of winning but for the sake of instruction.[7]

In the first decades after its introduction, the progress of the Reiswitz game was decidedly uneven. Funds were made available for the procurement of the game by each Prussian regiment. We know of the existence of wargame clubs

3 On German professional wargames, mission tactics, and the context of demographic growth and technological change, see also Bucholz, *Moltke, Schlieffen*.

4 The exception was Captain (later Colonel) Edmund Edler von Mayer; he served not in the Prussian but in the Austrian (later Austro-Hungarian) army; see also Meckel, *Studien*, 5–22; Zipser, *Anleitung*, 7; Hausrath, *Venture Simulation*, 20.

5 Meckel, *Studien*, distinguishes tactical, large, and strategic wargames; see also Meckel, *Anleitung*, 7–8; Mayer, *Studie*, 12–14; Zipser, *Anleitung*, 8–9.

6 See Mayer, *Studie*, 12–14; Meckel, *Studien*, 41–5; Meckel, *Anleitung*, 7–8; Zipser, *Anleitung*, 8–9.

7 Mayer, *Studie*, 3; repeated by Zipser, *Anleitung* 1; see also Anonymous, 'Anleitung zum Kriegsspiele', 65.

formed by Prussian officers in Berlin and other garrison towns. This suggests that the serious character of these games did not necessarily preclude playing for fun. A journal devoted to the game was published, but was not widely read.[8] In his *Geschichte des 24sten Infanterie-Regiments* ('History of the 24th Infantry Regiment'), Captain Franz von Zychlinski informs us that the game, which after all had been introduced by the King himself, was stored as a precious gift by the regiment commander, who held it in almost equal esteem as the regimental flag. He only allowed the game to be played when all officers of the regiment participated, but the available room was too small. Zychlinski tells how during one game, most junior officers left the cramped room to have refreshments in the kitchen. One of the arbiters, an older captain, started a premature and noisy evaluation of the game while the battle was actually still raging, in the process knocking down and dislocating some of the pieces. When the ensuing confusion had been dissolved, one of the senior commanding officers decided to confide the direction of his army to a younger officer, preferring to go out and have a drink. When he returned from this mission, he discovered that his subaltern colleague had bungled the whole operation, to the great frustration of the drinking absentee and to the even greater jollity of his colleagues. The resulting uproar ended the career of the game in the Twenty-Fourth Infantry Regiment.[9] It comes as no surprise that the game was reported out of sale in 1846. It seems that neither the stretch of peace between 1815 and 1864, nor the political conservatism of the period 1815–1848, nor the political reaction after the failed revolution of 1848 were suitable for the use of a novel military training device. It wasn't until the start of the 'New Era' (*Neue Ära*) with King Wilhelm, Chancellor Otto von Bismarck, and Minister of War Albrecht Theodor Emil von Roon, that we see substantial changes in foreign and military policy. These changes contributed to the victories in the Wars of German Unification (1864–1870).[10]

It is only around 1860 that professional wargames gained a firmer foothold—not only in the Great General Staff, but also in the field army.[11] In the 1870s the actual use of wargames became generally accepted in Germany, and also

8 Altrock, *Das Kriegsspiel*, 164; Young, *A Survey*, 21–2.

9 Zychlinski, *Geschichte des 24sten Infanterie-Regiments*, 196–9.

10 On post-Napoleonic reaction see Clark, *Iron Kingdom*, 399–408; on the 1859–1860 army reform see Messerschmidt, *Handbuch*, 160–217; against the idea of a military rupture with the time before 1859–1860, however, see Walter, *Preußische Heeresreformen*, 52–3.

11 Dannhauer, review of Verdy, 1063; Anonymous, *Anleitung zur Darstellung militärischer Manöver*, iii; Bucholz, *Moltke, Schlieffen*, 30; see also Leistenschneider, *Auftragstaktik*, 40.

increasingly amongst other nations.[12] Wargames fully partook in the flurry of military publications after the wars of 1864–1870, and came to be widely praised as useful devices to further military leadership.[13] The impression of stagnation between the defeat of Napoleon and the start of the 1860s, followed by a substantial increase in the 1860s, and a dramatic spike in the 1870s, is confirmed when we consult the bibliography of wargames in *Das Kriegsspiel* (1980) by Constantin von Altrock (1861–1942), see Table 5. In subsequent decades, wargames during times of peace could ruin careers of officers almost the way an actual war could. In 1890 Emperor Wilhelm II handed out such harsh criticism to a wargame solution provided by General Waldersee for a War Academy examination, that the general offered his resignation as head of the Great General Staff. This was not accepted. But when Waldersee in his turn criticized the Emperor's solution to a wargame problem in full hearing of a roomful of officers, the general was removed from the Staff and appointed commander of the 9th Corps in Altoona.[14]

The increased output of wargame literature in the 1860s and, especially, the 1870s was part of a military culture that permeated Germany during and after the string of spectacular military victories of the Wars of Unification.[15] The output of military texts not only included manuals for wargames themselves, but also showed a growing dissatisfaction with the conventional form of the game.[16] The deficiencies of the Reiswitz game type were listed most systematically by First Lieutenant (eventually General) Klemens Wilhelm Jacob Meckel (1842–1905). He was a veteran from the Franco-Prussian war and he became a member of the Great General Staff. In 1884 he was sent to Tokyo where he became an advisor to the Imperial Japanese Army General Staff between 1885 and 1888. His proposals for reform of the Japanese army were accepted by the Emperor. Back in Germany he would compose the opera 'Teja' and the Christmas oratorio 'Emperor Redbeard'. After his death in 1906 the Japanese military academy in Tokyo honoured him with a bust in its garden.

In his *Studien über das Kriegsspiel* (1873, 'Studies on the wargame'), Meckel observed that the decisions of the umpire were caught up in a mechanical and artificial pattern (*Schema*) of rules.[17] The rules did not help the umpire

12 Trotha, review of Verdy, 102; Anonymous, 'Anleitung zum Kriegsspiele', 66; Zipser, *Anleitung* 3.

13 Nauman, *Regiments-Kriegsspiel* vii; see also Leistenschneider, *Auftragstaktik* 10; Bucholz, *Moltke, Schlieffen*, 85.

14 Bucholz, *Moltke, Schlieffen*, 108.

15 See Wette, *Militarismus*, 48–64.

16 Bucholz, *Moltke, Schlieffen*, 85.

17 Meckel, *Studien*, 19, 24–5, 27, 47, 59.

TABLE 5 Publications about wargames in German (including reprints)

Decade	Number of publications about wargames
1801–1810	3
1811–1820	3
1821–1830	6
1831–1840	1
1841–1850	2
1851–1860	0
1861–1870	11
1871–1880	31
1881–1890	7
1891–1900	11

SOURCE: CONSTANTIN VON ALTROCK, *DAS KRIEGSSPIEL: EINE ANLEITUNG ZU SEINER HANDHABUNG. MIT BEISPIELEN UND LÖSUNGEN* (BERLIN: E.S. MITTLER, 1908), 160–90

to impart his military views to the players, but rather formed an impediment, making the role of the umpire unattractive. The rules of the game failed to do justice to the particular colour and character of each combat situation and gave the exercise a needlessly artificial character. The detailed calculation of losses with combat results tables made the game too slow, without substantially adding to the realism of its outcomes.[18] Moreover, games were often directed by uninterested or incapable umpires. These episodes of uninspired direction exacerbated mindless and mechanical styles of playing, something to which the game was already predisposed.[19]

Meckel did not only offer criticism; he (and other officers) tried to present solutions as well. He remarks that if art consists in the autonomous realisation of an idea, then warfare in the hands of a capable officer is an art.[20] He also notes that in the first decades of the nineteenth century military textbooks were based on general philosophical principles, presented in an abstract way. Whereas this earlier military literature had tried to grasp warfare in terms of

18 Meckel, *Studien*, 24–5.

19 Meckel, *Anleitung*, 59; see also Trotha, *Anleitung*, 15; Altrock, *Das Kriegsspiel*, 164; Young, *A Survey*, 21.

20 Meckel, *Studien*, 6; see also Meckel, *Studien*, 13; Trotha, *Anleitung*, 1.

a theoretical science, contemporary literature emphasized its character of a practical art. Hence Meckel considered the wargame umpire as a practical artist, or, as he put it, as 'an officer with the gift of independent action and phantasy'.[21] The umpire-as-artist is not needlessly impeded by the rules of the game and uses them in such a way that he can do justice to each particular situation.

The challenge for Meckel was to formulate rules for a wargame as a form of art that would not restrict a creative umpire with mechanical limitations.[22] The crux of the problem, as he saw it, was the dice and the combat resolution tables in conventional wargames. Meckel gives the example of a game situation where three infantry platoons (red) attack two infantry platoons (blue). Blue is placed in a favourable defensive position, while red attacks across an open field without cover. By attacking, red exposes itself to blue's gunfire. In a traditional game, even if red has lost one-third of its forces by the time it reaches the entrenched position of blue, its numerical equality at this point implies that it still has a good chance of defeating blue. This, Meckel pointed out, is a completely unrealistic scenario. Red's morale will suffer long before it has lost one third of its men during the attack and subsequent success against blue is very unlikely. So, the traditional game rules were unrealistic. Their mechanical character presented a young officer with 'a world of illusions' that were bound to cause 'disillusionment and despondency' as soon as they were confronted with the reality of battle.[23]

What made conventional wargames especially unsuitable in Meckel's view, was that a complete operation from start to finish, such as the attack by red against blue in the above example, was expressed by one single mechanical calculation. This single calculation did not take into account highly relevant intermediate steps. To address this problem, in his 'Studies' Meckel proposed a step-by-step procedure, wherein an active umpire who takes stock after each round of combat, considers the physical and moral conditions for that specific round, and then decides the odds for a favourable throw of the dice by picking the relevant column in the relevant combat results table. In the case of a deteriorating morale caused by steeply increasing losses during a hazardous attack, the umpire could decrease the odds for a favourable dice result from 'medium' to 'small' and subsequently even to 'smallest'. This would preclude the highly unlikely situation in which the remnants of three shattered attacking red platoons overcome two well-entrenched blue platoons. In this way the umpire had the chance to play the game as a true artist, with a sensitivity

21 Meckel, *Studien*, 27.

22 Meckel, *Studien*, 28; see also Naumann, *Regiments-Kriegsspiel*, vii, 12.

23 Meckel, *Studien*, 34.

for the shifting tides of battle, and with the power to translate this feeling (*Fingerspitzengefühl*) to facts.[24]

Meckel's idea to increase the realism of wargames through a more sophisticated use of dice and combat results tables by the umpire can also be found in the works of other contemporary game designers.[25] All these games may have gone some way in overcoming complaints about the overly mechanical and hence unrealistic character of the earlier Reiswitz game. But this was not the only complaint raised by Meckel: he also cautioned that the detailed calculation of losses with combat results tables made the game too slow. This latter complaint indeed seems to have accompanied the Reiswitz game right from its inception and was corroborated by its fate in the hands of the Twenty-Fourth Infantry Regiment. However, it is very hard to see how the innovations of Meckel made gameplay faster and easier; and it is very easy to see how the step-by-step use of combat results tables actually increased the cumbersome and tedious character of previous wargames.[26]

So how could professional wargaming be made more realistic than Reiswitz's game (which Meckel may very well have achieved), without sacrificing simplicity and hence playability (which Meckel almost certainly failed to do)? A surprisingly simple answer was formulated by Julius von Verdy du Vernois (1832–1910). In 1867 he was appointed head of the intelligence section of the Great General Staff, in which capacity he served throughout the Franco-German war. He rose to the rank of Major General and from 1889 to 1890 he served as Minister of War. After his retirement, he continued work on his military writings and he also produced the theatre play 'Alaric, King of the Visigoths'. He was extremely interested in the education of Germany's officer corps, and this is the background for his radical wargame innovations.[27]

While Meckel had allowed the umpire large measures of discretion in the use of dice and combat results tables, Verdy stopped using dice, tables, and detailed rules entirely. He modelled this simplification on the staff rides (*Übungsreisen*) of the German army.[28] Staff rides were exercises in which officers in the field conducted imaginary operations with imaginary troops over actual ground. All orders and messages were written out and submitted

24 Meckel, *Studien*, 34–6; see also Meckel, *Studien*, 20.

25 Tschischwitz, *Anleitung*, 1, 25; Trotha, *Anleitung*, 1, 13; Neumann, *Directiven*, 1, 18–21; Mayer, *Studie*, 43–60.

26 Young, *A Survey*, 21; Dannhauer, review of Verdy, 1063.

27 Schlegel, 'Verdy', 71–5.

28 Verdy, *Beitrag*, vi; see also Dannhauer, review of Verdy, 1065; Bucholz, *Moltke, Schlieffen*, 85; Hall, 'Battlefield Tours', 93–101.

to an umpire. Rides were held on different hierarchical levels by different service arms. The rides of the Great General Staff simulated operations of entire army corps.[29] Verdy was well acquainted with staff rides and in the introduction to his *Beitrag zum Kriegsspiel* (1876, 'Contribution towards the wargame') he writes how it had occurred to him that, when actual terrain was replaced by general staff maps, the simple principles used for staff rides could also be used for wargames. In staff rides, no use was made of fixed rules, tables, or dice. Rather, the umpire decided the outcome of battles as he saw fit—but he was supposed to give arguments for his decisions during the subsequent evaluation.[30] Verdy presented a wargame that used this same 'abridged' (*abgekürzte*) method, hoping it would please 'especially those esteemed comrades who until now may have been daunted by the combat resolution tables and the rules' of earlier wargames.[31]

Verdy's abridged game in many ways remained within the broad format established by Reiswitz. The game was played by two teams in separate rooms under the direction of an umpire who had a general map. At the start of each game the umpire formulated a general situation (*General-Idee*) for the two opponents, followed by a more specific mission statement (*Aufgabe*) for each of the participating units in that game.[32] Contrary to Reiswitz's game, however, Verdy's game was played on real staff maps; and while previous wargames had used special game pieces, Verdy's game could be played with simple matches.[33] The umpire used a measuring rod and a ruler to determine movement distances. The two opponents were allowed to push forward their units simulteneously until the moment when the umpire decided that they had come within sight of each other, at which point play proceeded turn by turn, based on the written orders of the commander of each party. In Verdy's wargame, the umpire had a very active role. Combat results depended solely on the umpire, who made flexible estimates based on the specific conditions.[34]

The text of Verdy's 'Contribution' does not provide a manual, but rather, presents an overview of a single mock game from start to finish—one based on a scenario invented by Verdy. The mock game starts as follows. On 1 August of an unspecified year, after an imaginary *Ost-Division* has been pushed back

29 Bucholz, *Moltke, Schlieffen*, 34, 103–4.

30 Verdy, *Beitrag* vi; see also Anonymous, review of Verdy, 722–7; Young, *A Survey*, 66; Bucholz, *Moltke, Schlieffen*, 85.

31 Bucholz, *Moltke, Schlieffen*, ix.

32 Verdy, *Beitrag*, 3–5.

33 Altrock, *Kriegsspiel*, 33.

34 Verdy, *Beitrag*, 13–68; Young, *A Survey*, 93–95.

by a stronger *West-Division* from Markranstäd to Leipzig, the outposts of the opponents face each other along the Elster river. On the evening of that day Zwenkau is occupied by a detachment of the West Division. The situation then zooms in from the level of two complete enemy *divisions* to the more detailed level of two enemy *detachments*. The East Detachment of the East Division receives information from its division headquarters about the occupation of Zwenkau by a West detachment on 2 August, 5 a.m. The mission of this East Detachment is to protect the division's left flank near Connewitz and to explore Zwenkau and its environments. For this mission the East Detachment is allocated two battalions of infantry, two squadrons of cavalry and one battery of artillery. The West Division plans to attack Leipzig on 2 August 9 a.m. from the direction of Lindenau. One of its detachments, the West Detachment facing the East Detachment of the East Division, receives the order to support this attack by advancing east of the Floßgraben canal and along the right bank of the Pleiße river. The West Detachment is allocated three infantry divisions and four squadrons of lancers for its mission. With this information the game starts. The leaders of the different elements of each detachment are played by different members of the participating teams of imaginary officers. Each step of each opponent and each decision of the umpire is minutely described; and each event description is followed by an analysis, often in counterfactual mode: had commander X done this, the result would have been that.[35]

Although Verdy's wargame was novel in its ruthless simplicity, he emphasized the continuity between his wargame and previous wargames. He claimed that the wargame club to which he belonged had followed similar procedural changes to the game nearly 20 years earlier.[36] In his review Thilo von Trotha also remarked that Verdy's game 'in essence matches the practice that had established itself in the older wargame'.[37] In a similar vein Meckel remarked that the Reiswitzian game type had been played *in spite* of rather than *thanks* to its specific rules.[38] Hence he was not surprised to note that already in the previous decades, players had indeed started to bend the otherwise inflexible rules to their liking. In Meckel's astute characterisation, players had started 'emancipating themselves from the rules'.[39] Moreover, in the preface to his *Anleitung*

---

35 Verdy, *Beitrag*, 23; on counterfactuals see above, chapter 5.

36 Verdy, *Beitrag*, viii.

37 Trotha, review of Verdy, 102.

38 Meckel, *Anleitung*, 6.

39 Meckel, *Anleitung*, 6; Anonymous, 'Anleitung zum Kriegsspiele', 66.

zum Kriegsspiele (1875, 'Instructions for the wargame'), Meckel explained that he had developed his own game at the explicit behest of Verdy.[40]

The old Reiswitz game type and the new Meckel game type made use of explicit rules, combat results tables, and dice, and would become known as 'rigid wargame' (*strenges Kriegsspiel*), while the alternative type established by Verdy would become known as 'free wargame' (*freies Kriegsspiel*).[41] Authors pointed out that the free wargame could only be played with very good umpires, whose role was much more vital than the role of umpires in rigid wargames; so while the free wargame might be suitable for the officers of the Great General Staff, regimental officers might be better served by a rigid wargame.[42] Although the rigid wargame thus continued to be played, Verdy's wargame gave its players and its arbiters exactly the kind of non-mechanical exercise in the practical art of warfare that the times seemed to demand, and it became a huge success.[43] Most subsequent German wargames were of the free (Verdy) type.[44] Free wargames were also implemented by the armies of other nations. The American Brigadier General Farrand Sayre (1861–1952) strongly believed that his compatriots should opt for the free rather than the rigid wargame, observing that 'rigid Kriegsspiel has been found by the Germans, who have given it a thorough trial, to be too great a strain upon the patience; and we have less patience than the Germans'.[45] Both the German and the Russian general staffs would end up using a free wargame to test encounters similar to the one that actually took place in East Prussia in the initial weeks of the First World War in 1914. Although the two staffs drew similar conclusions from their gaming, the German victory at the Battle of Tannenberg (under Verdy's pupil Paul von Hindenburg) has been ascribed, amongst other things, to the fact that the German army based its preparations on these wargame lessons, while the Russian army failed to do the same.[46] The use of wargames continued during the Second World War. The attack on Poland, Operation Barbarossa, and the

40 Meckel, preface in *Anleitung*, [i–ii]; see also Meckel, *Studien* 38–9; Dannhauer, review of Verdy, 1063, 1067.

41 Young, *A Survey*, 23–6.

42 Meckel, *Anleitung*, 36; Naumann, *Regiments-Kriegsspiel*, vii; see also Verdy, *Beitrag* vi.

43 Altrock, *Kriegsspiel*, 166–71. Verdy was not the first to publish a free wargame; this honour belongs to Anonymous, *Elementar-Begriffe vom Kriege*.

44 Young, *A Survey*, 65–6, 104.

45 Sayre, *Map Maneuvres*, 15.

46 Hausrath, *Venture Simulation*, 23–5; Schlegel, 'Verdy', 74.

Ardennes Offensive were all prepared by wargames.[47] Moreover, wargames were not only used for *prior* testing, but even *during* campaigns.[48]

3 Wargames and Mission Tactics

The free wargames of Verdy and its rigid predecessors served as practical case studies: they allowed senior officers to propose a mission and left the competing teams the freedom and opportunity to choose the means to accomplish this mission.[49] Viewed from this perspective of freedom of means to reach fixed ends, the most relevant context for Prussian wargames is formed by the Prussian military leadership concept of mission tactics (*Auftragstaktik*). For a good introduction to this concept we can again turn to the work of Verdy. He published a multi-volume *Studien über Truppenführung* ('Studies on Troop Leadership') in 1873–1874, shortly before he published his wargame (the 'Contribution') in 1876.[50] If war is an art that consists in the autonomous realization of an idea, then officers should be able to swiftly and flexibly translate the intentions of their superiors into successful action under capricious circumstances, e.g. changes due to interaction with the enemy, as well as topographical, meteorological, and logistical considerations. Accomplishing the mission in accordance with the aims of the commander is more important than using correct rules and fixed procedures. This holds true even if the rules for successful warfare could be assessed in the first place. Given the constant friction of the inherent dynamics of armed conflict it is never clear 'whether one has to take hold to the rules or their exception'.[51] And although warfare indeed seems to be subject to certain rules, the whole point of the military art consists in the ability to weigh the importance of different rules

47 Pias, *Computer*, 225–6, 244–5; see also Young, *A Survey*, 17–85; Hausrath, *Venture Simulation*, 5–8; Hohrath, 'Prolegomena', 146–52; Perla, *Wargaming*, 30 ff.; Creveld, *Wargames*, 151–6.

48 See Creveld, *Wargames*, 162: 'As an episode of November 1944 shows, the Germans maintained their faith in the method to the end. By an extraordinary coincidence, the US Army attacked the front of the German Commander-in-Chief West, Field Marshall Walter Model, in the midst of a wargame designed to simulate just such an offensive. Instead of breaking off the game, Model made it go on. The relevant orders were fed straight to the front, thus saving precious time'. See also Pias, *Computer*, 245; Köstlbauer, 'Strange Attraction' 175–8.

49 See also Mayer, *Studie*, 14–16; Hausrath, *Venture Simulation*, 9.

50 I use a later version: Julius Verdy du Vernois, *Studien über Truppenführung* (Berlin: E.S. Mittler, 1898).

51 Verdy, *Studien*, 7, 14.

according to different circumstances. Commanders should order *what* mission has to be accomplished, not *how* exactly this should be accomplished.[52] In his *Auftragtragstaktik im preußisch-deutschen Heer 1871–1914* ('Mission tactics in the Prusso-German Army 1871–1914') Stephan Leistenschneider explains that mission tactics allowed and even anticipated that officers would disobey orders—given the appropriate circumstances.[53] At the same time they were supposed to realize the *intentions* of their superiors (*Gefechtszweck*) with the strictest discipline (*mit strengsten Gehorsam*).[54]

The general *concept* of mission tactics was developed in the course of the nineteenth century and was not strictly limited to the tactical levels, but also embraced higher operational levels.[55] The concept was already applied before it became part of an official doctrine; and it became part of a doctrine a few years before the term itself was coined. Leistenschneider observes that mission tactics were applied on an improvised basis in the Wars of Unification and especially during the Franco-German War of 1870–1871, while the concept was developed more systematically in the years after the Wars of Unification.[56] In 1888 a new training and exercise regulation was introduced that embraced and incorporated the principles of mission tactics.[57] The *term Auftragstaktik* (as opposed to the older *concept*) was most likely coined in 1892 by Albrecht von Bugoslawski (1834–1905), who used the term in a derogatory sense in contrast to the traditional *Normaltaktik* that he himself preferred.[58]

Analysis by the German army of the initial use of mission tactics in the Franco-German War led to mixed conclusions. On the one hand it was appreciated that an increased autonomy of commanders at all levels of the chain of command could be very useful. On the other hand it transpired that this autonomy could have potentially dangerous and chaotic disintegrative side-effects.[59] When the later Chief of the Great General Staff Alfred von Schlieffen looked back at the military events of 1870 he was shocked by 'the unauthorised action and indiscriminate behaviour of lower commanders'.[60] These observations found their way into the *Exerzir-Reglement* of 1888, which not

52 Leistenschneider, *Auftragstaktik*, 90.
53 Leistenschneider, *Auftragstaktik*, 42; see also Wittmann, *Auftragstaktik*, 39.
54 Leistenschneider, *Auftragstaktik*, 83–4.
55 Oetting, *Auftragstaktik*, 108–9.
56 Leistenschneider, *Auftragstaktik* 21, 43–55; see also Oetting, *Auftragstaktik*, 90. See below, §7.4 for more on the improvised character of early mission tactics.
57 Anonymous, *Exerzir-Reglement*.
58 See Leistenschneider, *Auftragstaktik*, 101.
59 Leistenschneider, *Auftrakstaktik*, 54.
60 Quoted in Leistenschneider, *Auftrakstaktik*, 51.

only contained the principles of mission tactics, but also stressed the need to counterbalance pernicious side-effects.[61] Mission tactics could only work when their centrifugal tendencies were countered by a thorough military education that instilled a set of cooperative core values.[62] German commanders at all levels (both the tactical and the operational levels) needed to share what Leistenschneider calls 'a certain homogeneity of thinking'.[63] So the slightly paradoxical upshot is that the more important the *autonomous* accomplishment of missions by individual officers became, the more important it became to engrain them with a *shared* mind set.[64]

The mental unity required for a successful application of mission tactics was installed by a shared didactic experience of joint problem solving. Officers were trained to reach decisions independently, communicate these decisions clearly to others, and lead their soldiers in the realisation of the intentions of their superiors and in cooperation with their fellow officers.[65] Both Verdy's 'Studies on Troop Leadership' of 1873–1874 and his 'Contribution to the wargame' of 1876 were didactic contributions to the challenges posed by mission tactics. The 'Studies on Troop Leadership' started with a portrayal of a mock campaign (though it was clearly inspired by the Austro-German War of 1866) at the level of an entire Prussian army corps, and was followed by the detailed mission of an infantry division. This portrayal thus prefigured the *General-Idee* and the *Aufgabe* ('mission') that I discussed in Verdy's wargame.[66] Officers were asked to stop reading after he had formulated a specific problem; to think of solutions themselves; and then to continue reading and check their solution with Verdy's own solution. An obvious drawback of this method, as Verdy remarked himself, was that the reader could access the solution before he had worked it out by himself.[67] In fact, this limitation of one-sided case studies in the 'Studies' of 1873–1874 may have stimulated Verdy to create his interactive wargame in the 'Contribution' of 1876, which obviously did not have the same problem. In this way, there is a clear and natural continuity between Verdy's 'Studies on Troop Leadership', understood as a *static* exercise, and his 'Contribution towards the wargame', understood as a *dynamic* exercise in mission tactics.[68]

61 Anonymous, *Exerzir-Reglement*, 108–9; see also Leistenschneider, *Auftrakstaktik* 21.
62 Verdy, *Studien*, 15.
63 Leistenschneider, *Auftrakstaktik*, 94–5.
64 Caemmerer, *Die Entwickelung*, 185; see also Bucholz, *Moltke, Schlieffen*, 53.
65 Bucholz, *Moltke, Schlieffen*, 80.
66 See above, §7.2.
67 Verdy, *Studien*, 1.
68 Meckel, *Studien*, 7.

Wargame designers pointed out that wargames, by their interactive character, modelled the unpredictable character of war, the absence of standard solutions, and the need for officers to use their own judgement. This made wargames eminently suitable as training tools for mission tactics.[69] As Verdy observes, wargames helped players to transform mere *knowledge* about the need for autonomous action into *capability* for performing such action.[70] And while this point was made about professional wargames in general (on both the tactical and the operational level), it seemed to favour free wargames in particular. Free autonomous judgement was better served by free rules than by rigid rules, and it is evident that in Germany the future belonged to free wargames.[71] An increase in officers that were able to lead a wargame led to a reduced need for detailed rules.[72]

The use of professional wargames in the context of mission tactics became institutionalized in the Prussian and later German Great General Staff under Helmut von Moltke (1800–1891), who was its Chief between 1857 and 1888. In his memorandum of 1868 on the conclusions drawn from the war against Austria and in his famous *Verordnungen für die höheren Truppenführer* (1869, 'Instructions for Large Unit Commanders'), he had already formulated the main concepts of mission tactics.[73] These works did not amount to a coherent military doctrine let alone a textbook, and they were written for a restricted circle. Nevertheless, Moltke used the Great General Staff as a didactic platform to steep an entire generation of German officers in mission tactics.[74] In conscious opposition to Napoleon, Moltke tried to educate a class of autonomous leaders rather than mere executors of his orders.[75] Wargames, together

69 Meckel, *Anleitung* 5–22; Mayer, *Studie* 14–15; Naumann, *Regiments-Kriegsspiel*, vii. The link between wargames and mission tactics was not only made by the game designers themselves, but has also been noted by other authors, e.g. Altrock, *Kriegsspiel*, 8–11; Bucholz, *Moltke and the German Wars*, 59; Groß, *Mythos und Wirklichkeit*, 40–1.

70 Verdy, *Studien*, 14.

71 Meckel, *Anleitung*, 6, 47, 59.

72 Meckel, *Anleitung* 6, 47, 59; see also Zipser, *Anleitung*, 1; Young, *A Survey*, 76: 'It may be speculated, however, that the lack of extensive use of data charts may have been due to the increasing ability and knowledge of the officers who practised the war games. With increased participation in the movement of troops in the field, or in actual wars, came increased broadening of the experience and background of military men, making more feasible a Free Kriegsspiel such as von Verdy advocated'.

73 Moltke, *Memoire*; Moltke, *Verordnungen*; on Moltke and the Great General Staff see Groß, *Mythos und Wirklichkeit* 29–60; on Moltke and games and mission tactics see also Kessel, *Moltke*, 428–430, 449; Bucholz, *Moltke, Schlieffen* 1–108; Bucholz, *Moltke and the German Wars*, 33–6, 156–9.

74 Leistenschneider, *Auftragstaktik*, 54–62.

75 Caemmerer, *Entwickelung*, 134.

with manoeuvres, staff rides, lectures, the study of topography, and military history, played an essential role in the curriculum provided to officers of the Great General Staff. Moreover, staff officers disseminated the concept of mission tactics when they rotated to field regiments, where similar media were used increasingly to educate field officers.[76] Moltke himself already played Reiswitz's wargame in 1828, when he was still a lieutenant.[77] In 1837, when stationed in Constantinople, he taught the game to the Ottoman commander Chosref Pasha.[78] And when he was Chief of Staff to the Fourth Prussian Corps, his Magdeburg Club ranked first in 1844—although the club of the Prussian Guard Artillery was also a force to be reckoned with.[79]

So far, we have seen how significant wargames were for mission tactics and how both were embraced by the Great General Staff under Moltke. Let us now zoom out to a wider context that is equally relevant for both wargames and mission tactics.

4 Wargames and Mission Tactics: Incubation, Rifles, and Railways

Prussia's crushing defeat during the twin Battles of Jena-Auerstedt in 1806 caused her extensive territorial losses, brought the remainder of the Kingdom in the French sphere of influence, and energized a reform movement that sought to regenerate the Prussian state (often against the suspicions of Napoleon or the vacillations of the Prussian King Frederic William III himself). The Prussian reform movement had a political, educational, social, and economic agenda, but in many ways these aspects were instrumental for a goal that was ultimately military: a victory that would end French domination.[80] Perhaps the most important military reformer was Gerhard von Scharnhorst.[81] In 1807 he became head of the Military Reorganisation Commission that undertook the difficult task of reforming an ossified Prussian army into an organisation that was able to withstand the torrent of Napoleonic warfare.[82] In this role he was instrumental in the creation of a permanent Great General

76 Caemmerer, *Entwickelung*, 207; Bucholz, *Moltke, Schlieffen*, 30, 32–38.
77 [Troschke], 'Zum Kriegsspiel', 276.
78 Bucholz, *Moltke, Schlieffen*, 30.
79 Trotha, *Anleitung*, v; Young, *A Survey*, 22; Bucholz, *Moltke, Schlieffen*, 30.
80 Leistenschneider, *Auftragstaktik*, 29; see also Walter, *Preußische Heeresreformen*, 235–324; and see also above, §6.3, about the need for Prussia to 'educate its way out of the danger'.
81 Lehmann, *Scharnhorst*, i, 83–4.
82 Hall, 'The Modern Model', 96–7.

Staff. Also, following Napoleonic custom, he oversaw the division of the amorphous Prussian army in distinct army corps, which in turn were divided into separate divisions, thus creating independent units that could be used for missions under autonomous direction. At the tactical level Scharnhorst inspired a *Vorläufige Instruction für die Uebung der Truppen* (1808, 'Preliminary instruction for the exercise of Troops') that created the formation of special skirmishers, who were supposed to act with more flexibility than other parts of the Prussian infantry. According to the 'Preliminary instruction', skirmishers should be trained in order to fulfil their missions, but *how* they were trained should be left to the discretion of their officers.[83] Finally, Scharnhorst was the founding father of the *Akademie für junge Offiziere der Infanterie und Kavallerie* (1801) ('Academy for young infantry and cavalry officers'), the direct precursor of the *Kriegsakademie* (1810), with a curriculum that stressed critical thinking rather than procedural knowledge. In this way, Scharnhorst's measures provided the organisational and institutional space for what at this stage was merely a proto-concept of mission tactics.[84] We have no proof that at this early stage he articulated views about the use of wargames for this proto-concept—even though Scharnhorst himself may have introduced wargames in his lessons on tactics and strategy.[85] Moreover, before wargames and mission tactics could really be integrated, Napoleon was already defeated (1815), and the enthusiasm for reform in Prussia plummeted. In the post-Napoleonic decades we see many wargames meeting a fate that is typical for devices that are introduced with official sponsorship, but without much subsequent understanding of their use: they ended up in a cupboard. Similarly, in this period we see the introduction of a training and exercise regulation for the infantry in 1847 that was 100 pages longer than the regulation adopted during the crisis of Napoleonic domination. This regulation focussed on mindless drills rather than real combat training, and hence it was a well-regulated step backwards viewed from the desiderata of mission tactics.[86]

---

83 *Vorläufige Instruction* not found; quoted in Lehmann, *Scharnhorst*, i 154; see also Stübing, *Scharnhorst*, 65–91.

84 Lehmann, *Scharnhorst*, i 217–19.

85 See Holborn, 'The Prusso-German School', 283. The game in question was not necessarily the one by the older Reiswitz; in a letter of 8 September 1801 to Knesebeck (Scharnhorst, *Private und dienstliche Schriften*, 3, 59–60), Scharnhorst recommended Georg Venturini (1772–1802) for the position of teacher at a military institute and by this time Venturini had already published the first version of his prolific wargame (1798), see also above chapter 6. Interestingly, the younger Reiswitz used the statistics in Scharnhorst's *Über die Wirkung* for the combat results tables for the version of his game published in 1824.

86 Leistenschneider, *Auftragstaktik*, 36–7.

While Scharnhorst primarily contributed to an early *institutional* environment, his pupil Clausewitz provided an early *conceptual* basis for the eventual success of mission tactics.[87] In *On War* he explained that military theory should have a practical character that teaches autonomous officers to think critically about how to connect specific means with specific ends.[88] Given the highly uncertain character of warfare, officers who show initiative and courage will be more successful than officers who try to follow rules and procedures. Given this context it is not difficult to see the relevance of Clausewitz's military thought for the subsequent development of the concept of mission tactics.[89] But here again, as in the case of Scharnhorst, we first observe a period of static incubation. The end of this period was again formed by the start of the 'New Era' around 1860. Between the posthumous publication of *On War* in 1832–1834 and the 1860s Clausewitz was increasingly mentioned but little read.[90] More specifically, it is difficult to detect his influence on wargames published before the 1860s. This is not surprising; we have already seen that Clausewitz had little to say about wargames themselves.[91] Yet, in the works of wargame designers from the 1870s onwards he becomes a clearly detectable influence. His observations about the complex psychological and interactive forces that make it difficult for any commander to micromanage a military campaign would later be used by Meckel and Verdy to underpin the relevance of mission tactics and hence wargames.[92] In addition, the idea of warfare as a practical art as formulated by Meckel has clearly Clausewitzian overtones.[93] In some cases we even find explicit references to Clausewitz, for instance when Meckel explains how wargames can help officers train their powers of swift and independent decision making in spite of various forms of friction.[94]

So, it was only during the Wars of Unification that *mission tactics* were actually applied, while the concept was developed more systematically in the subsequent decade; and it was only from this time onwards that *wargames* became a widespread training device. Let us now concentrate on the technological context for this twin development. This context was of a much more massive

87 Gat, *Military Thought*, 151.
88 See above §5.4 and §6.5.
89 For Clausewitz's influence on Moltke see Caemmerer, *Entwickelung*, 68; Strachan, *Clausewitz's On War*, 10–14; Sigg, *Der Unterfüher*, 47–53.
90 Hahlweg, 'Das Clausewitzbild', 55.
91 See above, §6.6.
92 Clausewitz, *On War*, 140/Clausewitz, *Vom Kriege*, ii.2, 289; Clausewitz, *On War*, 149/Clausewitz, *Vom Kriege*, ii.3, 303; see also above §§4.4–4.5.
93 See above, §5.4.
94 Meckel, *Studien*, 13–14; see also Trotha, *Anleitung*, 1–2; Bucholz, *Moltke, Schlieffen*, 85.

and brutal nature than either Scharnhorst's institutional work or Clausewitz's conceptual work.

One very important context was formed by the dramatic increase in fire power in the course of the nineteenth century. Comparing rifles at the end of the nineteenth century with rifles used during the Napoleonic wars, Rudolf Karl Fritz von Caemmerer observed how their reach had increased fivefold, how their fire frequencies had tripled, and how the rapidity and accuracy of artillery fire had increased as well.[95] Various innovations contributed to this effect, including the replacement of muzzle-loaded firearms by breach-loaded firearms.[96] When German infantry soldiers attacked in conventional rigid columns in the war of 1870–1871, they were mowed down by the murderous fire of French *chassepot* rifles. Cavalry charges against firing infantry did not fare much better.[97] The German army was taught the harsh lesson that its soldiers should disperse in the face increased firepower, making a flexible use of the possibilities of the terrain under the autonomous initiative of local commanders.[98] This is how the actual *practice* of mission tactics, in addition to just its *concept*, was brought to life in deadly combat fire; and this is why mission tactics had the improvised character mentioned earlier.[99]

Since the dramatic effect of increased fire power during the German Wars stimulated mission tactics, which in its turn stimulated the use of wargames, it is not surprising that we see specific efforts to include the effects of fire power in wargames in the 1870s, when German game designers started to digest the lessons from spectacular victories—victories that had not only surprised Europe but also the Germans themselves.[100] When Meckel advocated a more realistic use of dice and combat tables with his step-by-step battle resolution approach, he very much tried to model the enormous and often non-linear impact of modern long-distance fire.[101] Was the firing unit placed in a favourable position? Had it recently rushed to its position or was it amply rested? Was it firing in an open or closed formation? What was the nature and the distance of the object that was subjected to fire? In addition to these physical considerations, Meckel also took into account psychological factors. Was the unit firing

95 Caemmerer, *Entwickelung*, 135–6; see also Verdy, *Studien*, 11.
96 Walter, *Preußische Heeresreform*, 130–7, 587–92.
97 Mayer, *Studie*, 72.
98 Leistenschneider, *Auftragstaktik*, 54, 73, 79–82; Sigg, *Der Unterführer*, 33–6.
99 See above, §7.3.
100 See Walter, *Preußische Heeresreform*, 18–22; Clark, *Iron Kingdom*, 551–2.
101 Meckel, *Studien*, 29–30.

under fire itself? Was it surprised by the fire of a swiftly approaching opponent? Was it suffering from increasing stress during a long exchange of fire?

Meckel tried to assign quantitative values to these factors, and the flood of military publications in the 1870s provided him with a treasure trove of statistical information about the performance of the latest firearms during the German wars. While these points had driven his *criteria* for an improved wargame in his 'Studies' of 1873, the 'Instructions' of 1875 was his attempt to produce a game that actually *applied* these criteria. The result was a complex combat resolution table as shown in Figure5, below. This table was designed to assess realistic combat outcomes in various specific situations, and its realism greatly surpassed attempts by designers of earlier wargames.[102] The effect of 2.5 minutes of fire by an infantry platoon is given for different targets at different distances, for example against artillery at a distance between 300 and 400 meters. In addition, before a dice is rolled to assess the losses of the targeted artillery, a choice must be made between five columns with different statistical bandwidths for different throws of the dice: smallest effect, small effect, medium effect, great effect, and greatest effect. In the case mentioned here, i.e. fire against artillery between 300 and 400 meters, the bandwidth for the smallest effect is 1 hit on a roll of 1, increasing to 5 hits on a roll of 6; while the greatest effect gives 30 hits on a roll of 1, increasing to 60 hits on a roll of 6.[103]

Increased firepower was not only considered something that should and could be modelled in a wargame, it was also used to provide legitimacy for using a professional wargame in the first place. Verdy noted that, given the destructive force of modern rifles and artillery, bravery without wits no longer sufficed. Increased fire power brought in its wake increased demands on the intellect of commanders.[104] And this is where his wargame could play a vital role. In this way, increased firepower stimulated the addition of the new medium of the wargame as a training device by proxy, in addition to the traditional medium of books on military history.[105] Moreover, in 1874 Edler von Mayer noted laconically that while *peace* does not provide an officer the chance to imagine the effect of fire, *war* often provides this insight too late, i.e.

102 See Mayer, *Studie*, 17; Bucholz, *Moltke, Schlieffen*, 87; see also above, §6.6.

103 Meckel was not the only game designer who tried to factor in the increased effect of fire power. When Edmund Edler von Mayer presented the rules for his new wargame in 1874, he also stressed the importance of the devastating effect of breach-loaded rifles. See Mayer, *Studie*, 19; see also Mayer, *Studie*, 3; Naumann, *Regiments-Kriegsspiel*, 40–1; but compare Mayer, *Studie*, 26.

104 Verdy, preface in *Studien*, [i–ii].

105 See also Mayer, *Studie*, 7.

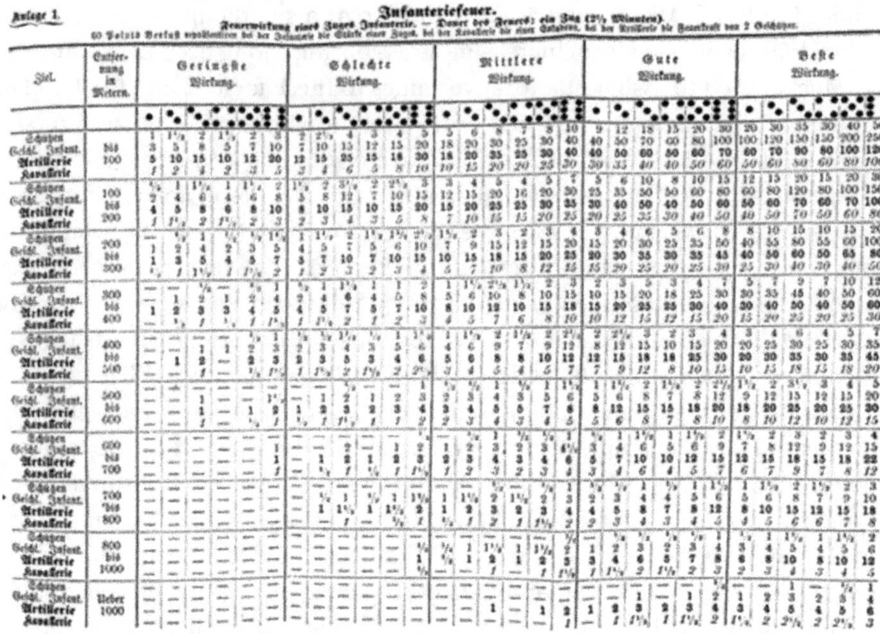

FIGURE 5 A detailed combat results table.
SOURCE: KLEMENS WILHELM JACOB MECKEL, *ANLEITUNG ZUM KRIEGSSPIELE I. THEIL DIREKTIVEN FÜR DAS KRIEGSSPIEL* (BERLIN: VOSSISCHE BUCHHANDLUNG, 1875), NO PAGE NUMBER PROVIDED

at the moment of dying; hence wargames might be a useful training device by proxy.[106]

So much for the importance of increased fire power for mission tactics and hence for wargames. Another context for the adoption of mission tactics in the 1860s was partly demographic and partly again technological; and here again we see a connection between mission tactics and wargames—more specifically, we see how wargames provided *solutions* for *challenges* created by mission tactics. In the decades after the Napoleonic wars the effect of population growth in Prussia meant that more youths were reaching military age. Accommodating these masses into the conservative structures of the Prussian army was an operation fraught with political problems and was at the heart of a constitutional crisis of 1862, but the results of military reform, pushed by demographic opportunities and international exigencies, were dramatic.[107] It

106 Mayer, *Studie*, 7.
107 Dupuy, *A Genius*, 51.

has been estimated that in the few years between 1864 and 1871, the annual number of soldiers mobilized by Prussia increased a staggering 1,500%, and that in terms of space used by the advancing army there was an increase of about 1,000%.[108]

Moreover, in the 1850s Germany crossed the dividing line between a pre-industrial and an industrial form of economy. Around 1850 the Prussian state started to actively sponsor and use railroads for military purposes, which may have quadrupled the speed of movement. Increased speed was achieved in particular during initial deployment. Once they arrived at the end of a rail line, military units reverted to traditional marching on foot.[109] Still, the combined result of demographic and technological change was a large army that could be transported swiftly over large distances. In addition, thanks to railroads the nation's vast sources of manpower, food, and equipment could be tapped. In the words of Rudolf von Caemmerer, railways made the country 'a single magazine with separate storerooms'.[110]

At the same time, these new possibilities brought new problems. Given the limited transport capability of any single railroad, the use of this fast mode of transport tended to favour movement in broad spaces. The individual corps of the army could only be amassed together for battle in the last stages of their movement; Moltke's concept 'March Divided, Fight United' was the fruit of these logistical challenges, and this motto was brilliantly brought into practice against the Austrians during the Battle of Sadowa in 1866 (see Figure 6).[111] In his 1869 'Instructions' Moltke clearly connects the motto with an increase of scale; and within a single page he draws a conclusion from this increase of scale that implies the use of mission tactics: 'There are many situations in which an officer should act in accordance with his insight. It would be very wrong to wait for orders when none can be given'.[112]

108 Bucholz, *Moltke, Schlieffen*, 54.

109 Creveld, *Supplying War*, 75–141.

110 Caemmerer, *Entwickelung*, 128; see also Walter, *Preußische Heeresreformen*, 124–8, 579–83; Peterson, 'A Game', 9.

111 Kessel, *Moltke*, 429–30.

112 Moltke, *Verordnungen*, 173, 174; see also Moltke, *Verordnungen*, 179–82; Bucholz, *Moltke, Schlieffen*, 41; Leistenscheider, *Auftragstaktik* 79. Moltke allowed large freedom for his corps commanders to achieve his aims. This *Führen durch Direktiven* ('leadership by directive') was also known as *Auftragsverfahren* ('misson method'); see Dupuy, *Genius for War*, 51–2; Bucholz, *Moltke, Schlieffen*, 56; Walter, *Preußische Heeresreformen*, 546–7. Moltke's own use of *Autragsverfahren* at the specific level of army corps was in line with his support for *Auftragstaktik* at all levels; see Moltke, *Verordnungen*, 180; Leistenschneider, *Auftragstaktik* 60; cf. Oetting, *Auftragstaktik*, 103–19.

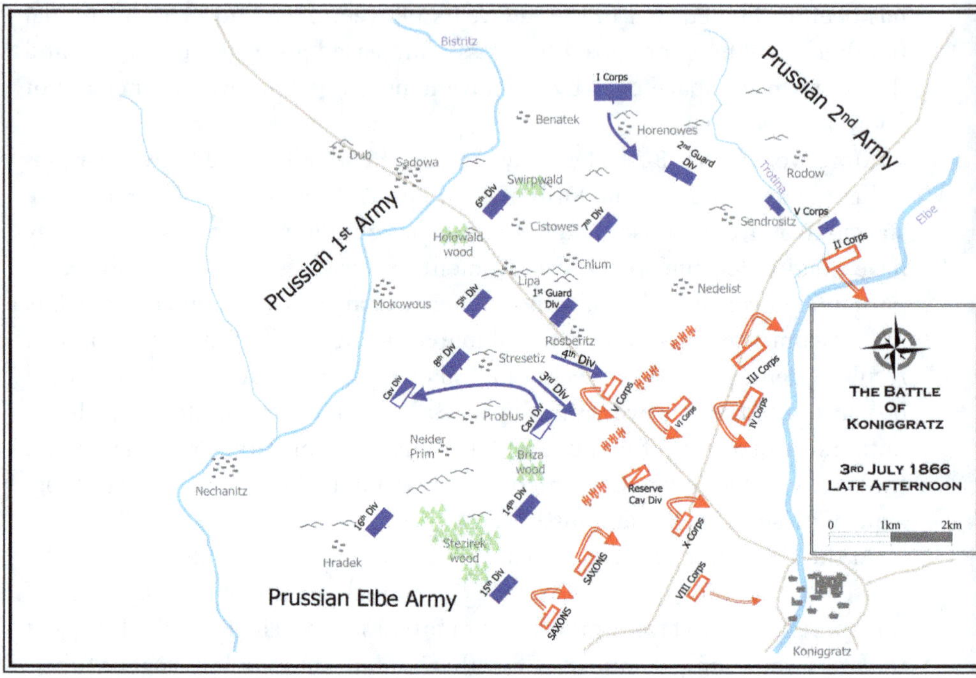

FIGURE 6 The Battle of Sadowa (or Königgrätz), 3 July 1866
SOURCE: HTTPS://WWW.PINTEREST.CO.UK/PIN/103231016441284541/, ACCESSED 27 OCTOBER 2020

The result of mechanized mass warfare with the help of railways had the same effect as increased fire power, i.e. spatial extension. Spatial extension caused by fire power had the form of dispersal on the tactical level of separate combats, while in the case of railway movement, extension took place at the operational level of entire army corps—and at this level extension was followed by concentration. At both levels of extension, the result was roughly the same: a reduced attention for procedures and an increased latitude for local commanders to choose the means that allowed them to accomplish the aims formulated by their superiors. Moreover, just as mission tactics stimulated by increased fire power at the *tactical level* demanded integrative devices, so did mechanized warfare at the *operational level*. Here again, wargames (along with staff rides and manoeuvres), played a vital integrative role. So, the need to instill a shared mindset through the use of wargames existed both at the tactical and at the operational level.

Finally, while the technological invention of railroads amplified the *disintegrative* tendencies of mission tactics, the invention of the telegraph clearly had an *integrative* effect. In 1904 Rudolf von Caemmerer would enthusiastically write that the telegraph 'has completely removed the dangers of [spatial] separation'.[113] Interestingly enough, though, the Great General Staff did *not* try to use the telegraph to micro-manage operations. The telegraph was in no way used to supplant either mission tactics or wargames. Actually, the Great General Staff took an active interest in integrating the telegraph in its mission tactics. The telegraph was not only used for swift communication between the Great General Staff and corps commanders, but also *between* corps commanders, who could thus continue to perform in the autonomous *modus operandi* demanded by mission tactics; and these cooperative operations needed *more* rather than *less* training by playing professional wargames.[114]

5 Conclusion

The first Prussian wargames were born in the aftermath of Prussia's crushing defeats at the hands of Napoleon in 1806. In the first decades of the nineteenth century Scharnhorst and Clausewitz also provided the institutional and conceptual space for the eventual development and introduction of mission tactics. But both wargames and mission tactics first went through a period of latency. Wargames were introduced by the state, but their actual use stagnated in post-Napoleonic Prussia. Similarly, the year 1847 saw the establishment of an exercise regulation that did not introduce mission tactics but moved away from that concept. A more complete concept of mission tactics was developed in Moltke's Great General Staff from the 1860s onwards, while its actual improvised introduction took place under the fire of French *chassepot* rifles in the war of 1870–1, which favoured dispersal of troops under fire. Dispersal favoured mission tactics, which carried disintegrative risks. These risks were countered by a host of integrative training devices, of which professional wargames were an essential part. Similarly, on the operational level, the use of railways for the swift deployment of entire army corps favoured initial dispersal (followed by subsequent concentration). Dispersal on this level again encouraged mission tactics, which again carried the risk of disintegration. This danger was again countered by wargames. The development of professional wargames and the proto-concept of mission tactics after Prussia's defeats in the Napoleonic wars initially had a precarious character. But once a firm link had been established between leading by mission tactics and training for leadership by wargames in the 1870s, mission tactics and wargames jointly spiralled upwards to increased

113 Caemmerer, *Entwickelung*, 133; see also Walter, *Preußische Heeresreformen*, 128–30, 583–6.
114 Caemmerer, *Entwickelung* 134, 137; see also Bucholz, *Moltke, Schlieffen*, 83.

levels of effectiveness. Constantin von Altrock was right when he boasted in 1908: 'Happily enough it can be established that on the fields of *military leadership* and *wargames* the German army doubtlessly marches ahead'.[115] Professional wargames and mission tactics were both *hesitant consequences* of Prussian defeat during the Napoleonic wars; they were both *substantial consequences* of the Prussian victories during the German Wars; and they may have been *significant joint contributors* to Germany's military successes during the First and Second World War.

115 Altrock, *Kriegsspiel*, iv (italics by me, P.S.); see also Bucholz, *Moltke, Schlieffen*, 1–2.

CHAPTER 8

Herbert Spencer and the Paradox of War

1 Introduction

War is a paradoxical phenomenon. The idea that a willingness to wage war can serve as a guarantee for peace has had adherents from Antiquity down to the present era; the Latin dictum *si vis pacem para bellum* ('If you want peace, prepare for war') and the observation that the post-1945 nuclear standoff between NATO and the Warsaw Pact directly contributed to an era of peace and prosperity in the Western world are all instances of the same paradox. In his fine *War, What Is It Good For?* (2014) Ian Morris posits an even stronger paradox: war itself, rather than merely a preparedness for war, has had massively benign effects for mankind. Morris claims that in spite of the horrendous immediate impact of war, in the long run 'productive' warfare put a premium on cooperation. This contributed to larger and more organized societies that protected their increasingly wealthy members from violence. Moreover, Morris claims that no other activity in history has been better at producing peaceful and prosperous societies than war.[1] He was not the first to articulate this strong paradox of war. In this chapter I will analyse an earlier and neglected version, formulated by the Victorian philosopher Herbert Spencer (1820–1903). He had a very keen eye for the on-going process of carnage and warfare 'among all sentient creatures' on both the biological and the human level.[2] His evolutionary sociology analyses the survival of the fittest societies in their struggle for existence.[3] The phrase 'survival of the fittest' was indeed coined by him and contributed to his image as a bloody-minded Social Darwinist.[4] But while many nineteenth-century philosophers crossed the border between military *is* and *ought* with breath-taking ease, Spencer was very careful with such steps.[5] Actually, his pacifist stance against contemporary forms of jingoism is well-documented. He often and eloquently depicts the horrors of war, including those perpetrated by the expanding British Empire, and especially

1 Morris, *War*, 7–9.

2 Spencer, *Biology*, 1, 340; Turner, *Herbert Spencer*, 76.

3 Spencer, *Sociology*, 2, 578.

4 Weikart, 'Darwin', 21–2.

5 Crook, *Darwinism*, 41; see also Battistelli, 'War and Militarism', 204; Renwick, 'Spencerian Science', 37.

 | DOI:10.1163/9789004536678_010

when conquest went hand-in-hand with religious hypocrisy.[6] Spencer was an extremely prolific writer and his theory of evolution amounts to a 'theory of everything'. Warfare looms large in his sociology, including his three-volume *Principles of Sociology* (1874–1896) and his *Study of Sociology* (1873); but his views on warfare can only be understood in connection with his *Principles of Biology* (1864–1867). Both his sociology and his biology in their turn are part of the broad canvas sketched in his *First Principles* (1862). A study of these latter works provides the context for Spencer's equally *clear* and *ambivalent* answer to the question 'War, what is it good for?' This question will be formulated and answered in terms of functionality: when according to Spencer is war functional, when is it dysfunctional, and what is the function in his sociology of this distinction itself? We will see how Spencer solved the riddle of the functionality of war with a highly original combination of an evolutionary and a system-theoretical approach.[7]

2 The Function of War: before Spencer

Spencer was not the first to think about the social functionality or dysfunctionality of war in a progressive, *stadial* context. Stadial models of history were informed by an Enlightened optimism about progress, peace, and science. An example is provided by Adam Ferguson (1723–1816) in his *Essay on the History of Civil Society*. Ferguson, with whose work Spencer was acquainted, noted that no activity is better suited to stimulate the activity and manly virtues that benefit the internal development of a society than military conflict with its neighbours.[8] In the savage and barbarian stages of history, the need for external violence stimulated peaceful internal cooperation and hence civilization. Moreover, the need for a vigorous spirit had remained in the more 'polished' societies at a later stage in history. If it has become increasingly difficult to gain this vigour in military battle, then it should be sought in a nation's internal market.[9] In Bernard Mandeville's model of private economic activity with

6 Spencer, *Sociology*, 3, 584; see also Spencer, *Sociology*, 2, 225; Spencer, *Sociology*, 3, 338; Turner, introduction in Spencer, *Sociology*, 1, lv; Bollender, 'Sociology', [28].

7 For a system-theoretical approach, see also above, §3.3 and §4.4; for Spencer's contribution to systems theory see Turner, introduction in Spencer, *Sociology*, 1, xiii; Ben-Gal, 'The Central Concepts', 192; Beetz, 'Das unliebsame System', 22–37.

8 Ferguson, *Civil Society*, 254, n. 97.

9 Ferguson, *Civil Society*, 105.

unintended beneficial consequences, Ferguson found a replacement for a function that had previously been fulfilled by warfare.[10]

Ferguson's analysis of a development in which warfare first provided positive functionality while becoming redundant in a later stage, was continued in the nineteenth century. In this century, the stadial model was often infused with biology as a paradigmatic science that was supposed to be instrumental in the development of an all-revealing 'science of society'.[11] Of special relevance here is Auguste Comte (1798–1857), whom Spencer started to read in 1851. Spencer would denounce Comtianism as unscientific, but Comte provided Spencer with his conception of the interaction between (social) organisms and their environment.[12] Moreover, Comte assigned a functionality to warfare in a stadial context, in which he, in his turn, was influenced by Claude-Henri de Rouvroy, count of Saint-Simon (1760–1820). Saint-Simon saw an all-important dichotomy between the feudal or theological system and the industrial or scientific system, between the feudal class and the industrial class, and between the feudal spirit and the industrial spirit. He used these categories to draw an optimistic historical development from the feudal to the industrial stage. Although Saint-Simon condemned war in the present, he appreciated its function in the past. He noted that in feudal times warfare had a clear function and all the parts of the political body were coordinated towards that purpose.[13] The feudal system remained functional as long as war itself remained the main purpose of a nation, but as the art of war became more perfect, it came to rely more on industry, and hence it gradually became more dependent on the industrial rather than the feudal classes. As the military power of nations came to depend less on drill and more on their industrial apparatus, it became increasingly difficult to wage war without disturbing economic production; hence warfare came to exert an increasingly deleterious influence on the whole of mankind, the victors included. So ultimately the increased industrial sophistication of warfare contributed to a marginalization of its functionality.[14]

Saint-Simon's dualism between war and industry can also be found in the work of his pupil Comte, who distinguished three stages in the evolution of human knowledge: the theological stage is associated with the military life, the positive stage is associated with the industrial life, while the metaphysical stage

---

10 Ferguson, *Civil Society*, 245; see also Ferguson, *Civil Society*, 210; Battistelli, 'War and Militarism', 197; and above, §2.8.

11 Crook, *Darwinism*, 36.

12 Guillo, 'Biology-inspired Sociology', 123–55; Pearce, 'From "Circumstances"', 246–50.

13 Saint-Simon, *Du système industriel*, x–xi.

14 Battistelli, 'Il problema', 3–15.

is merely an intermediate and transitory stage. In the theological stage, warfare stimulated the earliest forms of technology and it also forged social cooperation as a defence against enemy groups. Conquest resulted in ever larger political unites and within these units division of labour was stimulated—but ultimately warfare would be replaced by the hegemony of a scientific mindset, the rule of Harmony, and a Religion of Humanity.[15]

3 Spencer on the Function of War

Wars are fought by societies, and in his massive *Principles of Sociology* Spencer explains that societies are evolving systems.[16] He uses the term 'system' or 'aggregate' for a set of interacting components that form a complex whole and that are in constant interaction with surroundings that can themselves consist of one or more systems.[17] He describes his social systems in terms of structures that carry out functions. The cooperative needs of these structures stimulate the growth of a regulating system. The most urgent and thus the oldest functions of this regulating system are of a defensive and offensive nature. In addition, social systems have 'sustentive' functions (economic production) and distributing functions (transport and communication).[18] Spencer observes how part of the output of a (social) system is used by this same system as new input, which results either in positive (self-reinforcing) or negative (self-correcting) feedback loops.[19] The result is a complex pattern of interactions between environmental factors that influence functions that influence structures, and structures that influence functions that influence environmental factors.[20] On the one hand, functional demands must have preceded structural answers, but on the other hand structures tend to reach a stage where their own 'sustention' becomes the primary thing and their function the secondary thing, so that structures can easily outlast functions.[21]

Spencer not only classifies societies according to the level of the integration of their regulative, sustentive, and distributing functions, but also according to

15 Comte, *Cours*, leçons 56–7, 6, 1–385; see also Aron, *La Société*, 9–17; Crook, *Darwinism*, 37–9.
16 See Turner, introduction in Spencer, *Sociology*, 1, xiii; Ben-Gal 'The Central Concepts', 192; Beetz, 'Das unliebsame System', 22–37.
17 Ben-Gal, 'The Central Concepts', 188–91.
18 Spencer, *Sociology*, 1, 498–548.
19 See Spencer, *Sociology*, 1, 633; Spencer, *Sociology*, 2, 478; see also above, §3.6 and §4.4.
20 Spencer, *Sociology*, 1, 538–40; see also Turner, *Spencer*, 107–9; Ben-Gal, 'The Central Concepts', 152.
21 Spencer, *Sociology*, 3, 518; see also Spencer, *Sociology*, 2, 429; cf. Thomson, *Spencer*, 115.

the prevalence of structures that favour conflict or welfare. He distinguishes the 'militant type', which favours war, and the 'industrial type', which favours peace. He leaves no doubt that the contrasts between their traits 'are among the most important with which Sociology has to deal'.[22] Societies can have characteristics of both types, but since war formed the prime motive for cooperation amongst humans, be it for defensive or offensive motives, the first societies to appear in human history were of the militant type.[23] Military functional demands influenced the structure of societies from their very origin and have continued to do so. Initially the political regulating system was merely an extension of the military regulating system. Military heads grew into political heads, and 'in proportion as militancy is chronic, the organization proper to an army becomes the organization proper to the whole society'.[24] In this first stage politics was a continuation of war with other means—not the other way round. Only when civilizations became more complex did political and military organization become dissociated, but the relation between military officers who acquired civil functions and political functionaries who were imbued with a military ethos remained close.[25] Spencer notes instances of the relation between successful 'militancy' and the strengthening of political control as recently as the career of Napoleon and the unification of the German Empire.[26]

Spencer gives an amazingly rich and detailed analysis of the profound influence of military regimentation on the evolution of societies. The needs of warfare constituted the main motive for taxation and the resulting coercive organization was the instrument by which even more taxes could be extracted.[27] Spencer also notes the primitive identity of military institutions with institutions for administering justice, and he observes 'how close is the kinship between the modes of dealing with external aggression and internal aggression'.[28] Both functions were originally performed by a body formed of chief, head men, and people. The history of consultative bodies is equally informed by warfare. At first these structures were nothing more than councils of war, consisting of open-air meetings of armed men that deliberated on

22 Spencer, *Sociology*, 1, 574.
23 Spencer, *Sociology*, 2, 241.
24 Spencer, *Sociology*, 3, 483.
25 Spencer, *Sociology*, 2, 587.
26 Spencer, *Sociology*, 2, 338.
27 Spencer, *Sociology*, 2, 567; see also Spencer, *Sociology*, 2, 488–9.
28 Spencer, *Sociology*, 2, 492.

military measures. Only subsequently were the actions of these bodies provided with a wider political scope.[29]

Not only does Spencer ascribe an enormous impact to warfare, like many of his predecessors he also notes the predominantly positive character of this influence. Effective cooperation within a social group demanded likeness amongst its units for more effective action against members of other groups, which further increased the cooperation within groups and made their members even more similar.[30] In this way internal development and external conflict had a mutually reinforcing character. Thanks to this process societies saw a decline in the aggression between their members. Strong and militarily successful empires not only brought in their wake the oppression of their members but they were also instrumental in spreading civilization: 'The evidence obliges us to admit that subjection to despots has been largely instrumental in advancing civilization. Induction and deduction alike prove this'.[31] The evolution of political, professional, and ecclesiastical institutions prompted by military demands all point in the same beneficent direction.[32] Actually, military cooperation was the chief cause of social integration in general.[33] War, conquest, annexation, and institutional evolution, all contributed to an upward spiralling process.[34] If the horrors of war were indeed capable of providing such tangibly positive results, then this provides us with a fine specimen of the strong paradox of war. Spencer was baffled by this paradox, which is not surprising, given his deeply pacifist sympathies.[35] Hence his use of the phrase 'the evidence obliges us to admit', mentioned above. Similarly, when he admits 'that abject submission of the weak to the strong, however unscrupulously enforced, has in some times and places been necessary', he adds: 'In brief, trustworthy interpretations of social arrangements imply an almost passionless consciousness'.[36] A pacifist needs clinical eyes before he is prepared to appreciate the positive effects of carnage.

---

29 Spencer, *Sociology*, 2, 398; for the importance of war in Spencer, see also Bollender, 'Spencer', [13]; Turner, *Spencer*, 76, n. 51; Turner, 'Returning', 537.

30 Spencer, *Sociology*, 2, 286; see also Spencer, *Sociology*, 2, 231.

31 Spencer, *Sociology*, 2, 361.

32 Spencer, *Sociology*, 3, 316.

33 Spencer, *Sociology*, 2, 286; see also Spencer, *Sociology*, 3, 316: 'Though, as fast as society advances, and especially as fast as the militant type yields place to the industrial type, a centralized and coercive control, political and ecclesiastical, becomes less needful, and plays a continually decreasing part in social evolution; yet the evidence compels us to admit that at first it was indispensable'.

34 See also Turner, introduction in Spencer, *Sociology*, 1, xliv.

35 See Laity, *British Peace Movement*, 95.

36 Spencer, *Sociology*, 2, 232.

Although Spencer tries to appreciate the functionality of war in dispassionate terms, he nevertheless puts a distinct and relatively early cap on the ultimate reach of the strong paradox. We have already observed his point that structures have a tendency to perpetuate themselves, long after their functions have become superfluous, or even detrimental.[37] Not all structures are liable to a process of ossification in the same degree, but military structures are especially vulnerable in this regard.[38] The regimentation and forced cooperation implied by the militant type implies a structure that strongly resists change.[39] Spencer provides an extensive anthropological and historical analysis of military functions that eventually result in rigid and unproductive customs and structures. For instance, war, which is a predominantly masculine affair, stands in the way of a greater equality of the sexes and necessitates hard physical labour by women (who have to replace their warring husbands), who consequently will be prevented from producing 'more and better offspring'.[40]

Moreover, warfare stimulates the widespread use of class-badges and forms of obeisance and address, which all signify subjection.[41] Ecclesiastical institutions tend to strengthen this subjection and hence these institutions flourish in the militant type. Indeed, the forced cooperation and the physical and intellectual submission that is maintained in the militant type have far-reaching and nefarious intellectual consequences. The militant society has subjects that are exposed to 'the conspicuous and perpetual experience of personal agency which the militant *régime* yields'.[42] This constant experience of personal agency in the form of a despotic sovereign makes his subjects susceptible to a concept of causality that is *personal* as well; i.e. cause and effect are associated with individual agencies, whether human or divine or, as frequently happened, a mixture of the two. The idea of *impersonal* (natural or scientific) causation is utterly alien to these subjects. This reinforces the fit between the militant society and its citizens even more.[43]

All these phenomena lead to the conclusion that in the evolution of societies war first makes a paradoxically positive contribution, while at a later stage the paradox is defused by the increasingly dysfunctional contribution of warfare: 'From war has been gained all that it had to give. The peopling of

37 See also Spencer, *Sociology*, 2, 254.
38 Spencer, *Sociology*, 2, 260.
39 Spencer, *Sociology*, 2, 576, 601.
40 Spencer, *Sociology*, 1, 769; Spencer, *Sociology*, 1, 739–43.
41 Spencer, *Sociology*, 2, 141; see also Spencer, *Sociology*, 2, 157, 177, 196.
42 Spencer, *Sociology*, 2, 600.
43 Spencer, *Sociology*, 2, 600.

the Earth by the more powerful and intelligent race, is a benefit in great measure achieved; and what remains to be done, calls for no other agency than the quiet pressure of a spreading industrial civilization on a barbarism that slowly dwindles'.[44] In system-theoretical terms this is the inevitable moment when a self-reinforcing positive feedback loop is corrected by a self-correcting negative feedback loop. At this stage further development depends on the cessation of military activities and on a dismantling of the ossified structures of the militant type. And this, according to Spencer, is indeed what has happened in the evolution of human societies. In spite of their rigid character, the old structures of the militant type lose ground, and '[w]hile the benefits achieved during the predatory period remain a permanent inheritance, the evils entailed by it will decrease and slowly die out'.[45] Where the positive influence of the strong paradox remains open-ended for Morris, this is firmly terminated in Spencer's stadial model.

With the relative decline of the militant type, a new social form, called 'the industrial type' unfolds. This change has ramifications across the whole range of a society. As the grip of the militant type on a society loosens, the personal rights and liberties of its citizens increase.[46] The importance of consultative bodies with their clearly military origin declines, while the weight of representative structures that represent the power of the citizens increases.[47] The ownership of land, which in the militant type had become concentrated in the hands of a single ruler or his vassals, becomes the communal property and eventually the individual property of citizens in the industrial type.[48] Although Spencer presents the outline of a process that is of a gradual nature, the militant type and the industrial type are juxtaposed with the absolute contrariety of two conceptual prototypes, or 'ideal forms'.[49] The contrast is expressed in antagonistic terms of destruction and peaceful labour, forced and voluntary cooperation, and aggressive egoism as an *intrinsic* element of the militant type and egoism that is merely *extrinsic* in industrial life.[50] A transition between these very different types is only possible as long as the functions and structures of the militant type have not become completely ossified and have

44 Spencer, *Sociology*, 2, 664.
45 Spencer, *Sociology*, 2, 242; see also Spencer, *Sociology*, vol. 2, 665.
46 Spencer, *Sociology*, 2, 419.
47 Spencer, *Sociology*, 2, 440.
48 Spencer, *Sociology*, 2, 556.
49 Spencer, *Sociology*, 2, 568; Spencer, *Sociology*, 2, 606.
50 Spencer, *Sociology*, 1, 12; Spencer, *Sociology*, 1, 663; Spencer, *Sociology*, 2, 637–40; Spencer, *Sociology*, 3, 163; Spencer, *Sociology*, 3, 492; Spencer, *Sociology*, 3, 591.

retained a minimum of plasticity.[51] Societies which lack this plasticity will simply disappear, 'if not by violence, then by a decline consequent on inability to compete with younger and more modifiable societies'.[52]

Even in the parts of Spencer's *Principles of Sociology* that deal with the peaceful industrial type, however, war retains a looming presence. A society can only concentrate on peaceful activities as long as it is not dragged into military pursuits.[53] Spencer observes that 'thus far no civilized or semi-civilized nation has fallen into circumstances making needless all social structures for resisting aggression'.[54] Hence it is not surprising that the industrial type is less clearly defined than the militant type. The industrial type 'has its trait so hidden by those of the still-dominant [militant] type, that its nature is nowhere more than very partially exemplified'.[55] Dissolution of the older type and evolution of the newer type prompt Spencer to observe a 'mingling of changes' and in his Appendix A to the *First Principles* he predicts that this 'jumbling' of the different regulative systems of the militant and the industrial type will continue in the future.[56]

The evolution of societies from the militant to the industrial type implies a general notion of progress. Spencer was inspired in his thinking about the concept of progress by Karl Ernst von Baer's (1792–1876) principle, according to which each organism develops from a state of homogeneity to a state of heterogeneity.[57] In his *First Principles* Spencer *widened the scope* and *refined the content* of this principle.

Firstly, he *widened its scope* by using it as the model for his whole system of universal progress that included astronomical, physical, biological, and social systems.[58] While he had at first used the word 'evolution' specifically to describe the growth of the embryo, he later used evolution for development on all these levels. Evolution as defined in these general terms was governed on all levels by the same abstract principle of the conservation of energy, or the 'persistence of force'.[59] His ambition to apply the same general and uniform principles to different levels prompted his use and development of

51 Spencer, *Sociology*, 1, 571.
52 Spencer, *Sociology*, 1, 589.
53 Spencer, *Sociology*, 2, 607; on social and institutional plasticity see above, §3.3.
54 Spencer, *Sociology*, 2, 615.
55 Spencer, *Sociology*, 2, 603.
56 Spencer, *Sociology*, 2, 646; Spencer, *First Principles*, 464.
57 Spencer, *First Principles*, 270–91; see also Ben-Gal, 'The Central Concepts', 183–4.
58 Spencer, *First Principles*, 163–78; see also Armon, 'Beyond Darwinism's Eclipse', 182.
59 Bowler, 'The Changing Meaning', 106.

system-theoretical concepts that can indeed be fruitfully applied to entities of a wildly varying nature.[60]

Secondly, Spencer *refined von Baers's principle* through the development of his notion of evolution itself: 'Evolution is an integration and concomitant dissipation of motion, during which the matter passes from an indefinite incoherent homogeneity to a definite, coherent heterogeneity; and during which the retained motion undergoes a parallel transformation'.[61] The resulting impressive universal perspective was already illustrated by countless examples from all the systemic levels in the *First Principles* itself, before it would be even further elaborated in Spencer's subsequent works. For example, the idea that the motion of a system tends to increasing integration and differentiation is illustrated with the example of 'the advance from the independent fighting of separate warriors to the combined fighting of regiments' that are given specialist functions.[62] So, the general context for Spencer's thinking about social progress and the transition from the militant to the industrial type of society is already firmly established in the *First Principles.* In the *Principles of Sociology* his progressive outlook would take him even beyond the industrial type, with a speculation about a type of society that is devoted to 'higher activities'. While in the militant type the individual existed for the benefit of the state, and while in the industrial type the state existed for the benefit of the individual, ultimately life would no longer be for work, but rather work would be for life.[63]

Spencer hoped that in the collaboration of the states forming the Concert of Europe it was possible to detect an 'end to the re-barbarization which is continually undoing civilization'.[64] His use of the word 're-barbarization' indicates that progress from the militant to the industrial type is by no means a linear process. Although the 'persistence of force' tends to result in progress on all the systemic levels in the long term, there is scope for partial regressions.[65] These regressions can be short and sharp but given the majestic framework of Spencer's universal system, the duration of these regressions can also be rather long and their impact rather deep. In his *Principles of Sociology* he uses various contemporary and historical examples to make the general sociological point 'that revived belligerent habits re-develop the militant type of structure'.[66] For

60 Ben-Gal, 'The Central Concepts', 196.
61 Spencer, *First Principles*, 321.
62 Spencer, *First Principles*, 310–11; see also Spencer, *First Principles*, 301.
63 Spencer, *Sociology*, 1, 575.
64 Spencer, *Sociology*, 3, 610.
65 Bowler, 'Evolution', 109.
66 Spencer, *Sociology*, 1, 579.

instance, he notes that twice in its history the early modern Dutch Republic, which had advanced in the direction of the industrial type as no nation before, relapsed to a monarchy (he means the regime of the Stadtholder) under the sharp 'reactive influence of war' in 1672 and 1747.[67] Britain provides Spencer with instances of longer-term fluctuations. In the period of the American War of Independence and the wars of the French Revolution and Napoleon, England regressed towards a society with reduced personal freedom. The peaceful period of 1815–1850 saw the country taking large strides towards the industrial type with 'a great liberalization of political arrangements'. In the period after 1850, i.e. Spencer's own time, he notes again a return towards the militant type, and the extension of centralized administration and compulsory regulation that tends to develop in its wake.[68] In the Third Edition of the *Principles of Sociology* (1898) he remarks that the possession of foreign colonies and the implementation of 'coercive legislation affecting men's lives' have increased steeply.[69] Although England's regression towards the militant type may have been a defensive reaction against France and the unified German Empire, 'with growth of armaments has gone growth of aggressiveness'.[70] This is in line with his repeated general observation that structures can assume aggressive functions that are different from the defensive functions that prompted their evolution in the first place.

The phenomenon of regression from the industrial to the militant type is merely the manifestation at a social level of the general phenomenon of fluctuation and equilibrium that can be found at all levels of the Spencerian system. In the *First Principles* he observes that when a finger draws the prong of a tuning-fork out of its state of equilibrium and then liberates the prong, the force exerted by the finger does not disappear, but persists: 'As much force as the finger exerts, so much opposing force arises among the cohering particles. Hence, when the prong is liberated, it is urged back by a force equal to that used in deflecting it' (minus the effect of giving motion to the air and transformation into heat).[71] In this way, 'the rhythm of motion' expresses the persistence of force. Moreover, the image of the decreasing amplitude of the prong suggests how fluctuation can be followed by renewed equilibrium. This rhythm is also present at the social level, and at this level war is amongst the first phenomena mentioned by Spencer: 'War, exhaustion, recoil—peace,

67 Spencer, *Sociology*, 1, 579–80; see also above, §1.8.
68 Spencer, *Sociology*, 2, 626; Spencer, *Sociology*, 1, 579; Spencer, *Sociology*, 3, 156.
69 Spencer, *Sociology*, 3, 600–2.
70 Spencer, *Sociology*, 3, 600–1.
71 Spencer, *First Principles*, 215.

prosperity, and renewed aggression:—see here the alternation as occurring among both savage and civilized peoples'.[72] In the *Principles of Sociology* this rhythm receives a more detailed treatment in terms of alternations between the military and the industrial type. Where the centralized political power characteristic of the militant type generates pressures for decentralization, the decentralized power that marks the industrial type creates pressures for centralization.[73] But although this provides us with a *description* of the relatively short-term oscillations between the militant and the industrial type in a long-term progressive framework, we still have not analysed the evolutionary mechanisms that Spencer uses to *explain* these movements. What is the nature of these causal mechanisms in general, and is there a difference between the mechanisms behind the evolution of *militant* functions and structures and the evolution of *industrial* functions and structures?

4 Biological Evolution: Two Mechanisms

Spencer's fundamental distinction between a militant and an industrial type of society was not new. We have seen that this distinction had already been developed by Saint-Simon and Comte. What made Spencer's sociology so important was his fusion between an evolutionary and a system-theoretical perspective. This perspective puts him in a position to compare the evolution of different kinds of systems on completely different levels. Trying to understand social systems, he observes that these systems have certain characteristics in common with organic systems that are not shared with inorganic systems such as, for instance, the solar system. In both organic and the social systems, evolution pertains to functions and structures, as he had already explained in his *First Principles*.[74] Both kinds of systems are marked by a progressive increase, differentiation, and mutual interrelation of their functions and structures, in a way that cannot be observed in inorganic aggregates.[75] In addition, the functions of organisms and societies are continued after the demise of their particular elements, which again does not apply to inorganic nature.[76] Yet Spencer takes care to emphasize that as soon as biology has fulfilled the function of suggesting to

72 Spencer, *First Principles*, 213; see also Battistelli, 'War and Militarism', 201.
73 See also Turner, introduction in Spencer, *Sociology*, 1, xxvii.
74 Spencer, *First Principles*, 314–20; Spencer, *Sociology*, 1, 36; Spencer, *Sociology*, 1, 435, 442; see also Lightman, 'The "History"', 19.
75 Spencer, *Sociology*, 1, 459–61.
76 Spencer, *Sociology*, 1, 457.

him social evolution in terms of functions and structures, it can be taken away, having served as 'mere scaffolding'.[77] Indeed in his essay *The Social Organism* Spencer stressed some fundamental differences between organisms and societies.[78] In 'The Relations of Biology, Psychology, and Sociology' he bitterly complained about his critics unwillingness to appreciate that his analogy only applied to *certain traits* of organisms and societies.[79] He did not mistake analogy for evidence.[80]

With these caveats in mind, we should examine Spencer's *Principles of Biology* (1864–1867) and his analysis of the driving forces of evolution, before we can turn to his discussion of the forces that drive the evolution of society in general and war in particular. We have seen that Spencer had already explained in the *First Principles* that the universal coexistence of antagonist forces implies the law of rhythm, which in its turn implies both the decomposition of every force into divergent forces and the ultimate establishment of a balance.[81] A system that is in balance with its environment is *equilibrated* or adapted. In the *Principles of Biology*, Spencer stresses that equilibration or adaptation has a direct and an indirect form. Direct equilibration occurs when a new 'incident force' in an organism's surroundings *directly* results in a new equilibrium, by the exercise of new functions by new structures.[82] When these new structures are passed on to the offspring of the organism in question, we have Lamarckian evolution by acquired characteristics. Spencer admits that to prove this transmission is 'for several reasons, comparatively difficult'.[83] At the same time he thinks it absurd to suppose that functionally acquired by modifications should *not* be transmitted: 'It involves a denial of the persistence of force to say that A may be changed into A^1, and may yet beget offspring exactly like those it would have begotten had it not been changed'.[84]

But this is not to say that *direct* equilibration is the only force that drives biological evolution. There is also the possibility of *indirect* equilibration: 'And by the continual destruction of the individuals that are least capable of maintaining their equilibria in the presence of this new incident force, there must eventually be arrived at an altered type completely in equilibrium'.[85] The word

77 Spencer, *Sociology*, 1, 592–3; see also Spencer, *Sociology*, 1, 36.
78 Spencer, *The Social Organism*, 272–6.
79 Spencer, 'The Relations of Biology', 163–71.
80 Ben-Gal, 'The Central Concepts', 134; see also Zammito, 'Teleology', 753.
81 Spencer, *First Principles*, 392; see also above, §8.3.
82 Spencer, *Biology*, 1, 444.
83 Spencer, *Biology*, 1, 244–5.
84 Spencer, *Biology*, 1, 256.
85 Spencer, *Biology*, 1, 444.

'eventually' indicates that in this case equilibration is not direct, but indirect. Here the process of acquiring adaptive functions and structures is spread out over numerous generations. And this, of course, is Darwinian evolution: 'This survival of the fittest, which I have here sought to express in mechanical terms, is that which Mr Darwin has called "natural selection, or the preservation of favoured races in the struggle for life"'.[86] In the *First Principles*, published in 1857, Spencer had discussed evolution, before the appearance of Darwin's *Origin of Species,* in 1859. On the biological level he had tended to favour Lamarckian evolution. By the time he published the *Principles of Biology* (1864–1867) however, he had reached a stage where he accorded an equal status to both kinds of equilibration.[87] His treatment of the two different kinds of equilibration is remarkably symmetrical. Part III, 'The evolution of life', contains a chapter called 'Direct equilibration', which is followed immediately by a chapter called 'Indirect equilibration'. Moreover, Spencer appreciates that short-term direct equilibration does not rule out long-term indirect equilibration; he observes that the two processes 'are in reality simultaneous'.[88] Actually, this fits Darwin's own position, who had always remained prepared to admit that, in addition to modification through natural selection, other mechanisms for the generation of variation might exist, including Lamarckian use-inheritance.[89] Only a decade after Darwin's death in 1882, empirical advances would configure these variations in genetic terms, discrediting Lamarckian use-heritance.[90] This delay in the rejection of direct equilibration may have offered a conceptual window to Spencer for his even-handed treatment of direct and indirect equilibration. Spencer's use of both Lamarckian and Darwinian evolution in a single system can be seen as a remarkable prefiguration of the recent success of Neo-Lamarckian environmental epigenetics side-by-side with Neo-Darwinian evolution and a subsequent call for a unified theory of evolution.[91]

Spencer's balanced treatment of two different kinds of equilibration not only puts his phrase 'survival of the fittest' into a wider perspective, but also raises the question of his *criterion* for preferring one kind of equilibration over the other when he tries to explain the individual phenomena of biological evolution. Firstly, as long as the functions and structures of organisms were

86 Spencer, *Biology,* 1, 444–5.
87 See also Burkhardt, 'Lamarck', 793–805; Offer, 'Social Change', 308–9.
88 Spencer, *Biology,* 1, 466.
89 Offer, 'Social Change', 309.
90 Offer, 'Social Change', 309.
91 Skinner, 'Environmental Epigenetics', 1296–302; but cf. Penny, 'Epigenetics', 1758–60; see also Turner, 'Models of "Natural" Selection', 535–8.

comparatively simple and passive, they had few possibilities for coordinating their actions and adjusting directly to their environments, so that indirect equilibration was relatively important: 'But along with the gradual evolution of organisms having some activity, there grows up a kind of equilibration that is relatively direct. In proportion as the activity increases, direct equilibration plays a more important part'.[92] Secondly, direct equilibration takes place under the continuous short-term effects of external factors on the individual organisms of a biological species. If, however, external changes are so swift and dramatic as to be fatal to the individual organism in question, then equilibration will take place in a longer term and not at the level of the individual organism, but at the level of the entire species.[93] So, depending on the circumstances, sometimes direct equilibration and sometimes indirect equilibration is Spencer's preferred explanatory mechanism for biological evolution. Active systems are more susceptible to direct than to indirect equilibration, while abrupt changes tend to favour indirect over direct equilibration. Let us now investigate the relevance of these criteria for social evolution in the militant and the industrial stage.

5 Evolution of the Militant and the Industrial Type: the Same Two Mechanisms

Spencer makes little use of the terms equilibration or adaptation in the *Principles of Sociology*, yet the biological concepts of direct and indirect equilibration are clearly present and they have a clear relevance for the social evolution of the militant and the industrial type. Most obviously, and also most notoriously, Spencer notes that the indirect equilibration brought about by the Darwinian struggle for life has a special relevance for the militant type of society. Natural selection and survival of the fittest are not only biological but also social phenomena. Only societies with the most efficient forms of military cooperation will survive, and Spencer typically notes that this is in accordance with 'the laws of evolution at large'.[94] Between tribal groups who vary in the degree to which they are prepared to retaliate when their members are killed, there will tend to be 'a survival of the unforgiving'.[95] During the relentless

92 Spencer, *Biology*, 1, 468–9; see also Renwick, 'Spencerian Science', 40; cf. Armon, 'Darwinism's Eclipse', 174–5.

93 Spencer, *Biology*, 1, 473; see also Spencer, *Biology*, 1, 442; Spencer, *Biology*, 1, 463.

94 Spencer, *Sociology*, 2, 278–80.

95 Spencer, *Sociology*, 2, 593; see also Spencer, *Sociology*, 3, 331.

struggle for survival among societies, the militant type will be favoured: 'The social type produced by survival of the fittest, will be one in which the fighting part includes all who can bear arms and be trusted with arms, while the remaining part serves simply as a permanent commissariat'.[96]

These various applications of *biological* indirect equilibration to *social* indirect equilibration, and the supposedly positive evaluation of (military) conflict, have earned Spencer the rather infamous sobriquet of 'social Darwinist'. The actual picture was more complicated. Firstly, we have already seen that he was actually a pacifist whose positive evaluation of war as the earliest form of human cooperation and hence of human civilization was formulated with an obvious reluctance. Secondly, the indirect equilibration implied by his 'social Darwinism' is not limited to physical aggression, and continues long after the militant type of society has been superseded by the industrial type, although Spencer does not elaborate on the precise form of the resulting bloodless struggle: 'when, the struggle for existence between societies by war having ceased, there remains only the industrial struggle for existence, the final survival and spread must be on the part of those societies which produce the largest number of the best individuals—individuals best adapted for life in the industrial state'.[97] Thirdly, we have already noticed Spencer's point made in the *Principles of Biology* that as organisms become more complicated and more active, indirect equilibration through Darwinian natural selection is increasingly superseded by direct equilibration through the transmission of acquired characteristics. This process is continued until natural selection *within* human societies (between human individuals) is almost completely replaced by direct equilibration through cultural transmission: habits and experiences are accumulated, and the result is passed on indefinitely from one generation to another.[98] This still leaves open the possibility of struggle *between* societies, but (see the previous point) competition between societies of the industrial type will tend to have an increasingly non-military character.

Although *indirect* equilibration remains a force to be reckoned with, not only on the biological but also on the social level, *direct* equilibration on the social level is an increasingly important mechanism in the course of human evolution. It is also a mechanism that is easy to understand; it consists of cultural transmission. But while the mechanism of direct equilibration on the social level is very clear, we have already noted Spencer's admission that the

96 Spencer, *Sociology*, 2, 570; see also Spencer, *Sociology*, 2, 578.
97 Spencer, *Sociology*, 2, 610.
98 Spencer, *Sociology*, 2, 404–14; Spencer, 'Social Organism', 272.

mechanism of direct equilibration on the biological level is very obscure.[99] Hence it is not surprising that in this case Spencer sees the paradigmatic potential of sociology for biology. In the *Principles of Biology* he notes: 'From the law of adaptive modification in societies, we may therefore hope to get a clue to the laws of adaptive modification in organisms'.[100]

In the later essay, 'The Relations of Biology, Psychology, and Sociology', Spencer would indeed repeat the importance of a reciprocal influence of biology and sociology when he defended himself against one-sided interpretations of 'the parallelism I have asserted between certain traits of individual organisms and certain traits of social organisms'.[101] Actually, rather than exploiting biology for sociology, or vice versa, Spencer was primarily interested in using the contributions of both disciplines for his overarching universal system.[102]

It has been noted that Spencer discussed Darwinian struggle in the context of his social theory 'in a somewhat chastened Lamarckian form', which might suggest a certain confusion; but although his use of the two evolutionary mechanisms to explain the phenomenon of war was rich and subtle, it was not diffuse.[103] He used the two different concepts of evolution in a distinct way in order to analyse the differences between the militant and the industrial type. The needs of warfare stimulated social cooperation, and cooperation started an evolutionary process of direct equilibration in which increasingly complicated functions and structures responded to increasingly complicated environmental needs in a process of cultural transmission. Given the primacy of military needs, it is not strange that the militant type of society evolved first. This militant type initially created a civilized space that *enabled* the gradual evolution of the industrial type; then *hindered* the further evolution of this industrial type; and finally will be *overcome* by the industrial type. During this entire process of direct equilibration there are rhythmic disruptions in the form of military disasters during which the fittest societies survive in a decreasingly bloody process of indirect equilibration. This picture of gradual *direct* equilibration of military institutions and cataclysmic *indirect* equilibration of entire societies fits Spencer's use of his criterion for either *direct* or *indirect* equilibration on the biological level as explained at the end of the previous section. This is the nature of his elegant 'parallelism' between biological and social functions and structures.

---

99 See above, §8.4.

100 Spencer, *Biology*, 1, 194.

101 Spencer, 'Relations of Biology', [4].

102 See also Crook, *Darwinism*, 41.

103 See Crook, *Darwinism*, 43.

6 The Function of the Militant/Industrial Typology

Although Spencer's progressive view of the peaceful social and economic evolution through *direct* equilibration away from the military type towards the industrial type allows for dramatic regressions as part of a process of *indirect* equilibration, and although societies actually consist of mixtures that have characteristics of both types, we have already seen that the two social types themselves remain utterly antagonistic.[104] This rigid conceptual antagonism perhaps shows itself most strikingly in Spencer's disparaging remarks about the possibility of a positive technological spin-off from military innovations. He agrees that the phenomenon has existed and admits that it still exists, but accords it a rather dismal future. He believes that technological innovation has contributed towards a development away from the militant type towards the industrial type. In proportion to the extent that peace prevails over war, the importance of military technological spin-offs will dwindle: 'And though from the early days when flint arrow-heads were chipped and clubs carved, down to present days when armour-plates a foot thick are rolled, the needs of defence and offence have urged on invention and mechanical skill; yet in our own generation steam-hammers, hydraulic rams, and multitudinous new appliances from locomotive to telephones, prove that industrial needs alone have come to furnish abundant pressure whereby, hereafter, the industrial arts will be further advanced'.[105] So, military spin-offs become an ever rarer phenomenon. Moreover, even where military spin-offs still occur or have occurred, Spencer stresses that 'the destroying activities have been antagonistic to the productive activities', and he believes that destruction has overshadowed production.[106] As an example he gives medieval France, where chronic warfare between the fifth and the tenth centuries repeatedly broke up industrial organization. In this way, by robbing the militant type of merit after its initial functionality, he maintains its antagonism with the industrial type. But why, actually, is he so keen to maintain this antagonism? What is the *function* of this antagonism?

For Spencer the state of peace associated with the industrial type is very much a function of his political liberalism. This liberalism informs his comparison between the functionality of organic and social systems. In both cases the systems acquire specialized 'seats of feeling' that process information and regulate the system in question. This development of centralized structures has been more limited in the cases of societies then in the case of organisms.

104 See above §8.2.
105 Spencer, *Sociology*, 2, 665.
106 Spencer, *Sociology*, 3, 366.

Societies do not have the same physical cohesion between their parts as organisms. They do not consist of separate organs, but of separate people. Whereas in organisms consciousness is concentrated, it is diffused in societies. There is no 'social sensorium'; each individual has its own sensorium. Spencer concludes that whereas the organs of an organism exist for the sake of that organism, a society exists for the sake of its members, not the other way around—and this observation forms the basis for his liberalism.[107] In a liberal and peaceful organization, coercive structures whose function had been the protection against other societies fade away; but the structures that provide individuals protection against each other will remain in existence.[108]

Spencer notes that in his own time, however, the industrial type and its concomitant liberalism are still accompanied everywhere by various admixtures of the militant type, with the ever-present danger of regressions away from the industrial type towards the militant type. While the natural political principle of the industrial type is liberalism, that of the militant type is socialism or communism. Some socialists 'though probably not many', actually realize that their ideal modes of living have already prevailed in the earliest societies, 'as well as among some of the civilized who have lagged behind'. A militant society rests on forced cooperation; this tends to place 'a premium on idleness' and fits a socialist mode of production.[109] The burdening of 'the better' for the benefit of 'the worse' must 'check the evolution of a higher and more adapted nature: the ultimate result being that a community by which this policy is pursued, will, other things being equal, fail in competition with a community which pursues the purely equitable policy, and will eventually disappear in the race of civilization'.[110] In the long run the combination of militarism and socialism is doomed. Societies of this type will gradually fall behind in terms of direct equilibration, or collapse suddenly 'like a house of cards' as the result of indirect equilibration. An instance is formed by the case 'of the ancient Peruvians before a handful of Spaniards'—the society of the Incas being *very* militant and *very* socialist.[111]

Spencer argues that more recent and even contemporary regressions from the industrial to the militant type have again been accompanied by regressions from liberalism to socialism.[112] He sees 1850 as the start of a period of

107 Spencer, *Sociology*, 1, 461–2.
108 Spencer, *Sociology*, 2, 660.
109 Spencer, *Sociology*, 3, 575.
110 Spencer, *Sociology*, 2, 663.
111 Spencer, *Sociology*, 3, 608.
112 Spencer, *Sociology*, 2, 605.

heightened international military tension and tirelessly points out examples of a concomitant extension of centralized administration and compulsory regulation in Britain and elsewhere. He considers laws against contagious diseases, social legislation against poverty, social housing projects, the distribution of gas and water, and nationalized railways as equally nefarious examples of militant regressions: 'Instead of extending the principle proper to the industrial type, of providing quick and costless remedies for injuries, minor as well as major, which citizens inflict on one another, legislators extend the principle of preventing them by inspections. The arrangements in mines, factories, ships, lodging houses, bakehouses, down even to water-closets in private dwellings, are prescribed by laws carried out by officials'.[113] The ultimate horror perhaps, for Spencer, is the use of public funds for public libraries whose quality is then inspected by authorities who are again financed with tax money. In each of these cases we see self-reinforcing mechanisms at work; compulsory charity stimulates 'improvidence' which in its turn stimulates even more compulsory charity. The clearest examples of a resurgent militarism and socialism, however, are not detected in Britain, but in France and, even more, Germany. In these countries strong military traditions have been matched by loud claims for various forms of state interventions by social reformers.[114] In Germany these demands were graciously met in 1882 by 'Prince Bismarck's scheme of State-socialism'.[115]

In his intellectual biography of Spencer, Mark Francis has argued that Spencer's anti-socialism was not primarily the product of a *laissez-faire* liberalism. Spencer's liberalism in the early *Social Statics* (1851) was centred on peace and altruism, and he was not averse to an active role of the state in order to realize justice and equity, including the nationalization of land, since humanity should not be degraded by begging the lords of the soil for 'room for the soles of their feet'.[116] This brand of liberalism would still inform the later essays in *The Man versus The State* (1884), which, moreover, teaches us more about the sources of Spencer's anti-socialism.[117] Francis observes that for Spencer this was merely 'a subset of his hostility to democratic politics; he believed that the working class would attempt to legislate advantages for themselves just as the upper classed had done'.[118] Spencer saw democracy in general and socialism in

113 Spencer, *Sociology*, 1, 584.
114 Spencer, *Sociology*, 2, 633–4.
115 Spencer, *Sociology*, 2, 590.
116 Quoted in Francis, *Herbert Spencer*, 254; see also Francis, *Herbert Spencer*, 311.
117 Spencer, *The Man versus The State*, 59–175.
118 Francis, *Herbert Spencer*, 316.

particular as yet another revival of a political theory that would obstruct 'remedies for injuries' and hinder the development of justice and peace. Socialism should hence be considered as an anti-liberal impediment to progress.[119]

So Spencer's rigid sociological distinction between the industrial and the militant type is deeply connected with his liberal and anti-militarist sympathies. The exclusive connection of war and socialism with the militant type and the equally exclusive connection of peace and liberalism with the industrial type keep them clearly apart and increase their functionality. The growth away from the militant type and towards the industrial type means that war, after its initial function in establishing human society, becomes increasingly dysfunctional. In that sense Spencer's model can be regarded as an alternative to forms of the strong paradox of war as espoused by Morris, in whose views 'productive' warfare tends to remain functional—and hence paradoxical—throughout human history. In Spencer's model the strong paradox of war ultimately evaporates during the evolution from the militant to the industrial type. There is, however, a price for this solution, and this price has again the form of a paradox.

We have already seen how Spencer attaches preciously little value to industrial development in the form of *military spin-offs*. In addition, it could be said that he is equally unappreciative of military development in the form of *industrial spin-offs*; i.e. he does not show much interest in the possibility that the fruits of industrialization could be used as a *means* towards the accomplishments of *military aims*—which would place industry back again in the instrumental role that it had so clearly in Saint-Simon's feudal system or Spencer's militant type. The intensification of warfare by industrial means would indeed show itself as a gruesomely clear possibility shortly after Spencer's death, with the outbreak of the First World War. In that way, every element of progress away from the militant to the industrial type provides the potential for equally strong regressions back to the militant type. This could be called the *industrial-militant paradox*; economic *progress* has the potential of increasing rather than decreasing the amplitude of Spencer's *regressions* from the industrial to the militant type. An appreciation of this paradox seems to have been within his grasp at least two times. Firstly, we have seen him observing the power of modern 'steam-hammers [and] hydraulic rams' and their application for the production of 'armour-plates a foot thick' without taking the step suggested here. Secondly, since he formulated an explicit relation between *strong militarization* and ultimate *military weakness* in the case of the Inca empire, he could

119 Francis, *Herbert Spencer*, 316.

but did not formulate a similar relation between *strong industrialization* and ultimate *military strength*. Actually, it is not surprising that Spencer did not appreciate the industrial-militant paradox, given its lethal ramifications for his model of social evolution. The industrial-militant paradox relates increased military potential with increased industrialization; but Spencer's militant-industrial model hinges on the assumption of increased pacification in the wake of increased industrialization. The process of industrialization was an ambiguous phenomenon; for Comte and Spencer it promised a pacific union of mankind, while for such different thinkers as Marx and Nietzsche its lethal weapons were unleashing an era of mass warfare.[120]

7 Evaluation

Spencer's lack of appreciation of the industrial-militant paradox may betray a certain awkwardness in the face of modern technological developments in his stunningly comprehensive evolution of everything—or as Fabrizio Battistelli put it: 'In its attempt to demonstrate the continuity of the process of evolution, from the Protozoa to the City of London, this approach is much more at home with the functions and structures of social aggregation in its simplest forms, than with those belonging to societies with a considerable level of historicity'.[121] Yet here Battistelli seems to miss the novelty of Spencer. He was not the first to use a stadial model of history in which warfare is given an initial but transitory function of war. We have seen this in Ferguson, Saint-Simon, and Comte.[122] Nevertheless, he was the first to introduce a functional–structural understanding of warfare, as part of his two theories of evolution, in the framework of a general systems theory. This framework allowed for a *naturalistic* analysis of war that is neither *vitalistic* nor completely *reductive*. Spencer's approach is *naturalistic* because he believes that rules of behaviour cannot be deliberately designed; they are the natural product of social life which have evolved gradually.[123] His approach is *anti-vitalistic* because of his belief that organisms and societies employ the same physical forces as inorganic nature—only when the organic level has thus been robbed of its uniqueness is it applied to the social level. This allows Spencer to posit the same Force at

120 Coker, 'The Collision', 66.
121 Battistelli, 'War and Militarism', 197.
122 Immanuel Kant could be added to this list, see Schuurman, 'Kant on War', 6–8.
123 Spencer, *Sociology*, 2, 216.

all system-theoretical levels.[124] Finally, his approach is *not* completely *reductive* because it leaves room for what can be called a complex systems model of teleology, in the sense that biological and social feedback systems can be described in terms of a goal-directedness whose aim is the survival or propagation of the system.[125] Although Spencer's evolutionary perspective has been the object of trenchant criticism, there is an appreciation that 'there are very few ideas employed within the mainstream of [modern] evolutionary sociology which cannot be traced back to his writings'.[126]

Compared to that of his predecessors, Spencer's system-theoretical and evolutionary analysis of war is original. Looking forward to his successors in the contemporary field of big history, his contribution remains surprisingly relevant. Historians such as Cynthia Stokes Brown, David Christian, and Fred Spier are all using the largest possible canvas for their analysis of the past, and their multi-level accounts are frequently inspired by system-theoretical and evolutionary concepts.[127] Since aggression, struggle, and war seem such universal phenomena, it is not surprising that some of the finest specimens of big history can be found in military history, in the works of William McNeill, Azar Gat, and Ian Morris.[128] The early contribution of Spencer's sociology to this field can best be understood in terms of a distinction made by Auguste Comte: that of a *statique sociale* whose objects are laws and the taxonomy of institutions, and that of a *dynamique sociale* that is interested in their evolution.[129] Thanks to Spencer's interest in the historical *dynamics* of the militant/industrial typology he was able to make his precocious contribution to big history—a contribution that is overlooked by most modern big historians.[130] Spencer was not interested in 'the gossip of history' but in a 'general law in the evolution of societies'.[131] He believed that societies are not constructed by designs and intentions but are the result of gradual natural processes. This belief was motivated by a liberal distrust of 'social schemers' and politicians

124 Spencer, *Sociology*, 1, 146.

125 See Zammito, 'Teleology', 752; Wattles, 'Teleology', 461; Christensen, 'A Complex Systems Theory', 305–8.

126 Offer, 'Social Change', 306.

127 Stokes Brown, *Big History*; Christian, *Maps of time*; Spier, *Big History*; see also Hawkey, 'A New Look', 163–79.

128 McNeill, *The Pursuit of Power*; Gat, *War*; Morris, *War*.

129 See Comte, *Cours*, leçon 48, 4, 151–247.

130 Cf. Spier, 'Big History: The Emergence', 143–6; Zakariya, 'Still a Fraud?', 636.

131 Spencer, *Sociology* vol. 3, 181.

who passed acts 'to amend acts that were before amended'.[132] It is within this general historical and ideological framework that Spencer first posits and then dissolves the strong paradox of war.

132 Spencer, *Sociology*, vol. 3, 324.

Bibliography

Algaier, Johann, *Der Anweisung zum Schachspiel zweyter Theil* (Wien: F. J. Rötzel, 1796).

Altrock, Constantin von, *Das Kriegsspiel: Eine Anleitung zu seiner Handhabung. Mit Beispielen und Lösungen* (Berlin: E.S. Mittler, 1908).

Anonymous, 'Supplement zu den bisherigen Kriegsspiel-Regeln', *Zeitschrift für Kunst, Wissenschaft und Geschichte des Krieges*, 13/4 (1828), 68–105.

Anonymous, *Elementar-Begriffe vom Kriege, durch Beispiele erläutert, mit einer Anleitung zu praktischen Uebungen für Militz-Offiziere in Form eines technischen Kriegsspieles* (Zürich: Schulthess, 1840).

Anonymous, *Anleitung zur Darstellung militärischer Manöver mit dem Apparat des Kriegs-Spiels* (Berlin: Berliner Kriegsspiel Verein, 1846).

Anonymous, 'Anleitung zum Kriegsspiele', *Organ der Militär-wissenschaftliche Vereine*, 13 (1876), 65–74.

Anonymous, review of Julius von Verdy du Vernois, *Beitrag zu den Kavallerie-Uebungsreisen* (Berlin: E.G. Mittler, 1876), in *Militair-Wochenblatt* (1876), 41, 721–9.

Aretin, Wilhelm Freiherr von, *Strategonon: Versuch die Kriegsführung durch ein Spiel anschaulich darzustellen* (Ansbach: Dollfuss, 1830).

Armon, Romy, 'Beyond Darwinism's Eclipse: Functional Evolution, Biochemical Recapitulation and Spencerian Emergence in the 1920s and 1930s', *Journal for the General Philosophy of Science*, 41 (2010), 173–94.

Armstrong, Aurelia, 'Natural and Unnatural Communities: Spinoza Beyond Hobbes', *British Journal for the History of Philosphy*, 17 (2009), 297–305.

Aron, Raymond, *Les étapes de la pensée sociologique : Montesquieu, Comte, Marx, Tocqueville, Durkheim, Pareto, Weber* (Paris: Gallimard, 1967).

Aron, Raymon, *La Société industrielle et la guerre: Suivi d'un Tableau de la diplomatie mondiale en 1958* (Paris: Librairie Plon, 1959).

Aron, Raymond, *Penser la guerre: Clausewitz*, 2 vols (Paris: Gallimard, 1976).

Baldwin, Jason, 'Hume's Knave and the Interests of Justice', *Journal for the History of Philosophy*, 42 (2004), 277–96.

Ball, Philip, *Critical Mass: How One Thing Leads to Another* (London: Arrow Books, 2005).

Barbon, Nicholas, *A Discourse of Trade* (London: Th. Milbourn, 1690).

Bassford, Christopher, Daniel Moran, and Gregory W. Pedlow, eds, *On Waterloo: Clausewitz, Wellington and the Campaign of 1815* (n. p., Clausewitz.com, 2010).

Battistelli, Fabrizio, 'Il problema della guerra dalla filosofia della storia alla sociologia: Claude-Henri de Saint-Simon', *Rassegna Italiana di Sociologia*, 28 (1986), 3–29.

Battistelli, Fabrizio, 'War and Militarism in the Thought of Herbert Spencer: With an Unpublished Letter on the Anglo-Boer War', *International Journal of Comparative Sociology*, 34 (1993), 192–209.

Baylis, John, and James J. Wirtz, 'Introduction: Strategy in the Contemporary World: Strategy after 9/11', in John Baylis, James J. Wirtz and Colin S. Gray, eds, *Strategy in the Contemporary World: An Introduction to Strategic Studies* (Oxford: Oxford University Press, 2013), 1–14.

Beckmann, Rasmus, *Clausewitz trifft Luhmann: Eine systemtheoretische Interpretation von Clausewitz' Hundlungstheorie* (Wiesbaden: Springer, 2011).

Beetz, Michael, 'Das unliebsame System: Herbert Spencers Werk als Prototyp einer Universaltheorie', *Zeitschrift für Soziologie*, 39 (2010), 22–37.

Bell, David A., *The First Total War: Napoleon's Europe and the Birth of Modern Warfare* (London: Bloomsbury, 2008).

Ben-Gal, Shraga, 'The Central Concepts Underlying Herbert Spencer's Evolutionary and Sociological Theories', PhD thesis, The Hebrew University of Jerusalem, 1980.

Benrekassa, Georges, *La politique et sa mémoire: Le politique et l'historique dans la pensée des Lumières* (Paris: Payot, 1983).

Béraud, Stéphane, *Le révolution militaire napoléonienne*, i: *Les manoeuvres* (n. p., B. Giovangeli, 2013).

Berry, Christopher, *The Idea of Luxury: A Conceptual and Historical Investigation* (Cambridge: Cambridge University Press, 1994).

Bertalanffy, Ludwig von, *General System Theory: Foundations, Development, Applications* (New York: G. Braziller, 1968).

Beyerchen, Alan, 'Clausewitz, Nonlinearity, and the Unpredictability of War', *International Security*, 17 (1992–1993), 59–90.

Blom, H.W., and I.W. Wildenberg, *Pieter de la Court en zijn tijd* (Amsterdam: Holland Universiteits Pers, 1986).

Bollender, Ronald Keith, 'The Sociology of Herbert Spencer', www.bolenderinitiatives.com/sociology/herbert-spencer-1820-1903/herbert-spencer-sociology-herbert-spencer, accessed 15 March 2015.

Bomel-Rainelli, Béatrice, 'Les rapports de Bossuet et de Montesquieu dans les manuels d'histoire de la littérature de 1841 à 1992', *Revue Montesquieu*, 5 (2001), 125–46.

Bonolas, Philippe, 'Fénelon et le luxe dans le *Télémaque*', *Studies on Voltaire and the eighteenth century*, 249 (1987), 81–90.

Bossuet, *Discours sur l'Histoire Universelle*, in B. Velat and Yvonne Champailler, eds, *Oeuvres* (Paris: Gallimard, 1961), iii, 659–1028.

Botterill, George, 'Two Kinds of Causal Explanation', *Theoria*, 76 (2010), 287–313.

Bowler, Peter J., 'The Changing Meaning of "Evolution"', *Journal of the History of Ideas*, 36 (1975), 95–114.

Brecke, Peter, 'Violent Conflicts 1400 A.D. to the Present in Different Regions of the World', paper for the 1999 Meeting of the Peace Science Society (8–10 October 1999), Ann Arbor, Michigan, http://www.inta.gatech.edu/peter/PSS99_paper.html, accessed 6 November 2010.

Bruin, Boudewijn de, 'Game Theory in Philosophy', *Topoi*, 24 (2005), 197–208.

Bucholz, Arden, *Moltke and the German Wars, 1864–1871* (Houndmills: Palgrave, 2001).

Bucholz, Arden, *Moltke, Schlieffen, and Prussian War Planning* (Houndmills: Palgrave, 2001).

Bunzl, Martin, 'Counterfactual History: A User's Guide', *The American Historical Review*, 109 (2004), 845–58.

Burke, Peter, 'Tacitism, Scepticism, and Reason', in J.H. Burns and Mark Goldie, eds, *The Cambridge History of Political Thought* (Cambridge: Cambridge University Press, 2006), 479–98.

Burkhardt Jr., Richard W., 'Lamarck, Evolution, and the Inheritance of Acquired Characters', *Genetics*, 194 (2013), 793–805.

Caemmerer, Rudolf Karl Fritz von, *Die Entwickelung der strategischen Wissenschaft im 19. Jahrhundert* (Berlin: W. Baensch, 1904).

Cahen, Albert, introduction to François de Fénelon, *Les Aventures de Télémaque*, ed. Albert Cahen, 2 vols (Paris: Hachette, 1920), 2, xv–cii.

Carrithers, David, 'Montesquieu's Philosophy of History', *Journal of the History of Ideas*, 47 (1986), 61–80.

Champblanc, Franz Dominik, *Das Kriegsspiel, oder das Schachspiel im Großen* (Wien: H.F. Müller, 1828).

Charrak, André, 'Le sens de la nécessité selon Montesquieu : Essai sur le live 1, chapitre I de *L'Esprit des lois*', *Revue de Métaphysicque et de Morale*, (2013), 7–17.

Christensen, Wayne. 'A Complex Systems Theory of Teleology', *Biology and Philosophy*, 11 (1996), 301–20.

Christian, David, *Maps of Time: An introduction to big history* (Berkeley: University of California Press, 2004).

Citino, Robert M., 'Military Histories Old and New: A Reintroduction', *American Historical Review*, 112 (2007), 1070–90.

Clark, Christopher, *Iron Kingdom. The Rise and Downfall of Prussia, 1600–1947* (London: Penguin, 2007).

Clausewitz, Carl von, *Vom Kriege*, 3 vols (Berlin: F. Dümmler, 1834).

Clausewitz, Carl von, *Vom Kriege*, ed. Werner Hahlweg (18th ed., Bonn: F. Dümmler, 1973).

Clausewitz, Carl von, *Leitfaden zur Bearbeitung der Taktik der Gefechtslehere*, in Werner Hahlweg, ed., *Vom Kriege* (18th ed., Bonn: F. Dümmler, 1973), 1103–80.

Clausewitz, Carl von, *Die wichtigsten Grundsätze des Kriegführens zur Ergänzung meines Unterrichts bei Sr. königlichen Hoheit dem Kronprinzen*, in Werner Hahlweg, ed., *Vom Kriege* (18th ed., Bonn: F. Dümmler, 1973), 1047–86.

Clausewitz, Carl von, *Bemerkungen über die reine und angewandte Strategie des Herrn von Bülow oder Kritik der darin enthaltenen Ansichten 1805*, in Werner Hahlweg, ed., *Verstreute kleine Schriften* (Osnabrück: Biblio Verlag, 1979), 65–88.

Clausewitz, Carl von, *Strategie aus dem Jahre 1804*, in Werner Hahlweg, ed., *Verstreute kleine Schriften* (Osnabrück: Biblio Verlag, 1979), 3–45.

Clausewitz, Carl von, *Strategie aus dem Jahre 1808*, in Werner Hahlweg, ed., *Verstreute kleine Schriften* (Osnabrück: Biblio Verlag, 1979), 50–61.

Clausewitz, Carl von, *On War*, ed. and trans. Michael Howard and Peter Paret (Princeton, NJ: Princeton University Press, 1984).

Clausewitz, Carl von, *Feldzug von 1815. Strategische Uebersicht des Feldzugs von 1815*, in Werner Hahlweg, ed., *Schriften — Aufsätze — Studien — Briefe* (Göttingen: Vandenhoeck & Ruprecht, 1990), ii, 2, 943–1118.

Clausewitz, Carl von, *Niederschriften des Werkes 'Vom Kriege'*, in Werner Hahlweg, ed., *Schriften — Aufsätze — Studien — Briefe* (Göttingen: Vandenhoeck & Ruprecht, 1990), ii, 1, 630–44 and ii, 2, 645–717.

Clausewitz, Carl von, *Ueber das Fortschreiten und den Stillstand der kriegerischen Begebenheiten*, in W. Hahlweg, ed., *Schriften — Aussätze — Studien — Briefe* (Göttingen: Vandenhoeck & Ruprecht, 1990), ii, 1, 248–55.

Clausewitz, Carl von, *On Wellington*, ed. and trans. Peter Hofschröer (Norman: University of Oklahoma Press, 2010).

Clausewitz, Carl von, *The Campaign of 1815*, in Christopher Bassford, Daniel Moran and Gregory W. Pedlow, eds and trans., *On Waterloo: Clausewitz, Wellington and the Campaign of 1815* (n. p., CLAUSEWITZ.COM, 2010), 55–218.

Clausewitz, Carl von, *On Waterloo: Clausewitz, Wellington and the Campaign of 1815*, ed. and trans. Christopher Bassford, Daniel Moran and Gregory W. Pedlow (n. p., CLAUSEWITZ.COM, 2010).

Coetzee, Daniel, and Lee W. Eysterlid, eds, *Philosophers of War: The Evolution of History's Greatest Military Thinkers*, 2 vols (Santa Barbara, California: Praeger, 2013).

Cohen, H. Floris, *How Modern Science Came into the World: Four Civilizations, One 17th-Century Breakthrough* (Amsterdam: Amsterdam University Press, 2012).

Coker, Christopher, *Barbarous Philosophers: Reflections on the Nature of War from Heraclitus to Heisenberg* (London: Hurst & Company, 2010).

Coker, Christopher, 'The Collision of Modern and Post-Modern War', in Julian Lindley-French and Yves Boyer, eds, *The Oxford Handbook of War* (Oxford: Oxford University Press, 2013), 57–68.

Colson, Bruno, 'Clausewitz on Waterloo', *War in History*, 19 (2012), 397–400.

Comte, Auguste, *Discours sur l'Esprit positif* (1842), http://classiques.uqac.ca/classiques/Comte_auguste/discours_esprit_positif/Discours_esprit_positif.pdf, accessed 12 November 2015.

Comte, Auguste, *Cours de philosophie positive*, 6 vols (Paris: Schleicher, 1907–1908).

Cormier, Youri, 'Hegel and Clausewitz: Convergence on Method, Divergence on Ethics', *The International History Review*, 36 (2014), 419–27.

Court, Johan de la, and Pieter de la Court, *Het Welvaren van Leiden: Handschrift uit het jaar 1659*, ed. F. Driessen (The Hague: Martinus Nijhoff, 1911).

Court, Johan de la, and Pieter de la Court, *Consideratien van Staat, ofte Politike Weegschaal* (4th ed., Amsterdam: Dirk Dirksz, 1662).

Court, Johan de la and Pieter de la Court, *Politike Discoursen, handelende in Ses onderscheid Boeken van Steeden, Landen, Oorlogen, Kerken, Regerringen en Zeeden* (2nd ed., Amsterdam: Ciprianus vander Gracht, 1662).

Court, Pieter de la, *Interest van Holland* (Amsterdam: J.C. vander Gracht, 1662).

Court, Pieter de la, *Aanwysing der heilsame politike gronden en maximen van de republike van Holland en West-Vriesland* (Leiden: Hakkens, 1669).

Court, Pieter de la, *The True Interest and Political Maxims of the Republick of Holland and West-Friesland* (London, 1702, no publisher given).

Court, Pieter de la, *Political Maxims of the State of Holland* (London: J. Nourse, 1743).

Courtois, Jean-Patrice, *Inflexions de la rationalité dans 'L'Esprit des Lois'; Écriture et pensée chez Montesquieu* (Paris: Presses Universitaires de France, 1999).

Cowley, Robert, ed., *What If? The World's Foremost Military Historians Imagine What Might Have Been* (London: Pan Books, 2001).

Cowley, Robert, ed., *What If? Eminent Historians Imagine What Might Have Been* (London: Pan Books, 2002).

Creveld, Martin van, *Supplying War: Logistics from Wallentstein to Patton* (Cambridge: Cambridge University Press, 2004).

Creveld, Martin van, *Wargames: From Gladiators to Gigabytes* (Cambridge: Cambridge University Press, 2013).

Crook, Paul, *Darwinism, War and History: The Debate over the Biology of War from the Origin of Species to the First World War* (Cambridge: Cambridge University Press, 1994).

Dannhauer, Ernst Heinrich, review of Julius Verdy du Vernois, *Beitrag zum Kriegsspiel* (Berlin: E.G. Mittler, 1876), in *Militair-Wochenblatt* (1876), 61, 1063–7.

Diniz, Eugenio, and Domício Proença Júnior, 'A Criterion for Settling Inconsistencies in Clausewitz's *On War*', *The Journal of Strategic Studies*, 37 (2012), 879–902.

Donker, Paul, 'The Genesis of Clausewitz's *On War* Reconsidered', *British Journal for Military History*, 2 (2016), 101–17.

Donker, Paul, 'The Evolution of Clausewitz's *Vom Kriege*: A Reconstruction on the Basis of Earlier Versions of His Masterpiece', https://www.clausewitzstudies.org/bibl/DonkerP-TheEvolutionOfClausewitzsVomKriege.pdf, accessed 30 October 2020.

Downing, Brian M., *The Military Revolution and Political Change. Origins of Democracy and Autocracy in Early Modern Europe* (Princeton, NJ: Princeton University Press, 1992).

Dunnigan, James F., *Wargames Handbook* (3rd. ed., San Jose: Writers Club Press, 2000).

Dupuy, T.N., *A Genius for War: The German Army and General Staff, 1807–1945* (Fairfax, Virginia: Hero Books, 1984).

Ehrard, Jena, '"Rome enfin que je hais ..."?', in Alberto Postigliola, ed., *Storia e ragione: Les Considérations sur les causes de la grandeur des Romains et de leur décadence di Montesquieu nel 250° della publicatione* (Naples: Liguori, 1987), 23–32.

Engberg-Pedersen, Anders, *Empire of Chance: The Napoleonic Wars and the Disorder of Things* (Cambridge, MA: Harvard University Press, 2015).

Evans, Richard J., *Altered Pasts: Counterfactuals in History* (Waltham, MA: Brandeis University Press, 2013).

Feld, Maury D., 'Middle-Class Society and the Rise of Military Professionalism: The Dutch Army 1589–1609', *Armed Forces and Society*, 1 (1975), 419–42.

Fénelon, François de, *Les Aventures de Télémaque*, 2 vols, ed. Albert Cahen (Paris: Hachette, 1920).

Fénelon, François de, *Avis de M. de Fénelon à une dame de qualité, sur l'éducation de mademoiselle sa fille*, in Jacques le Brun, ed., *Oeuvres*, 2 vols (Paris: Gallimard, 1983, 1997), ii, 1132.

Fénelon, François de, *Dialogues des morts composés pour l'éducation d'un prince*, in Jacques le Brun, ed., *Oeuvres*, 2 vols (Paris: Gallimard, 1983, 1997), i, 279–510.

Fénelon, François de, *Examen de conscience sur les devoirs de la royauté*, in Jacques le Brun, ed., *Oeuvres*, 2 vols (Paris: Gallimard, 1983, 1997), ii, 973–1012.

Fénelon, François de, *Examen des droits de Philippe V à la couronne d'Espagne*, in Jacques le Brun, ed., *Oeuvres*, 2 vols (Paris: Gallimard, 1983, 1997), ii, 1068–73.

Fénelon, François de, *Explication des maximes des saints*, in Jacques le Brun, ed., *Oeuvres*, 2 vols (Paris: Gallimard, 1983, 1997), i, 1001–98.

Fénelon, François de, *Instruction gérérale pour avoir la paix intérieure*, in Jacques le Brun, ed., *Oeuvres*, 2 vols (Paris: Gallimard, 1983, 1997), i, 656–671.

Fénelon, François de, *Lettre à Louis XIV*, in Jacques le Brun, ed., *Oeuvres*, 2 vols (Paris: Gallimard, 1983, 1997), i, 541–54.

Fénelon, François de, *Mémoire sur la campagne de 1712*, in Jacques le Brun, ed., *Oeuvres*, 2 vols (Paris: Gallimard, 1983, 1997), ii, 1074–7.

Fénelon, François de, *Mémoire sur les moyens de prévenir la guerre de la succession d'Espagne*, in Jacques le Brun, ed., *Oeuvres*, 2 vols (Paris: Gallimard, 1983, 1997), ii, 1013–27.

Fénelon, François de, *Mémoire sur la situation déplorable de la France en 1710*, in Jacques le Brun, ed., *Oeuvres*, 2 vols (Paris: Gallimard, 1983, 1997), ii, 1034–44.

Fénelon, François de, *Sur le pur amour*, in Jacques le Brun, ed., *Oeuvres*, 2 vols (Paris: Gallimard, 1983, 1997), i, 656–671.

Fénelon, François de, *Oeuvres*, ed. Jacques le Brun, 2 vols (Paris: Gallimard, 1983, 1997).

Fénelon, François de, *Tables de Chaulnes*, in Jacques le Brun, ed., *Oeuvres*, 2 vols (Paris: Gallimard, 1983, 1997), ii, 1085–105.

Fénelon, François de, *Telemachus, son of Ulysses*, ed. and trans. Patrick Riley (Cambridge: Cambridge University Press, 1994).

Ferguson, Adam, *An Essay on the History of Civil Society*, ed. Fania Oz-Salzberger. (Cambridge: Cambridge University Press, 2007).

Ferguson, Niall, ed., *Virtual History: Alternatives and Counterfactuals* (New York: Basic Books, 1997).

Ferreyrolles, G., 'Histoire et causalité chez Bossuet', in G. Ferreyrolles, ed., *Bossuet: Le Verbe et l'Histoire (1704–2004)*, Actes du colloque international de Paris et Meaux pour le troisième centenaire de la mort de Bossuet (Paris: H. Champion, 2006), 185–96.

Firmas-Périès, Armand Charles Daniel, *Le jeu de stratégie ou les échecs militaires* (Memmingen: J. Rehm, 1807).

Force, Pierre, *Self-interest before Adam Smith: A Genealogy of Economic Science* (Cambridge: Cambridge University Press, 2003).

Francis, Mark, *Herbert Spencer and the Invention of Modern Life* (Stocksfield: Acumen, 2007).

Freedman, Lawrence, *Strategy: A History* (Oxford: Oxford University Press, 2013).

Fukuyama, Francis, 'The End of History', *The National Interest* (Summer 1989), 3–18.

Fukuyama, Francis, *The End of History and the Last Man* (London: Penguin, 1992).

Gallagher, Catherine, 'The Formalism of Military History', *Representations*, 104 (2008), 23–33.

Gat, Azar, *A History of Military Thought from the Enlightenment to the Cold War* (Oxford: Oxford University Press, 2001).

Gat, Azar, *War in Human Civilization* (Oxford: Oxford University Press, 2006).

Gilroy, James P., 'Peace and the pursuit of happiness in the French utopian novel: Fénelon's *Télémaque* and Prévost's *Cleveland*', *Studies on Voltaire and the Eighteenth Century*, 176 (1979), 169–187.

Girard, René, *Battling to the End: Conversations with Benoît Chantre* (East Lansing: Michigan State University Press, 2010).

Golec, Piotr, 'Feedback', http://www.economicswebinstitute.org/glossary/feedback.htm, accessed 21 August 2020.

Goodman, Nelson, *Fact, Fiction and Forecast* (Cambridge, MA: Harvard University Press, 1983).

Goyard-Fabre, Simone, 'Le destin de Rome et la nature des choses', in Alberto Postigliola, ed., *Storia e ragione: Les Considérations sur les causes de la grandeur des Romains et de leur décadence di Montesquieu nel 250° della publicatione* (Naples: Liguori, 1987), 113–38.

Goyard-Fabre, Simone, *Montesquieu: Nature, les Lois, la Liberté* (Paris: Presses Universitaires de France, 1993).

Groß, Gerhard, *Mythos und Wirklichkeit: Geschichte des operativen Denkens im deutschen Heer von Moltke d. Ä. bis Heusinger* (Paderborn: F. Schöningh, 2012).

Guillo, Dominique, 'Biology-inspired Sociology of the Nineteenth Century: A Science of Social "Organization"', *Revue Française de Sociologie*, 43 (2002), 123–55.

Hahlweg, Werner, 'Das Clausewitzbild einst und jetzt', in Carl von Clausewitz, *Vom Kriege* (Bonn: F. Dümmler, 1973), 1–172.

Hall, David Ian, 'The Modern Model of the Battlefield Tour and Staff Ride: Post-1815 Prussian and German Traditions', *The Quarterly Journal*, (2002), 93–102.

Hall, David Ian, 'Battlefield Tours and Staff Rides: Prussian and German Traditions', *Connections: The Quarterly Journal of the Partnership for Peace Consortium of Defence Academics and Security Studies Institutes*, 1 (2005), 93–101.

Hampton, Jean, *Hobbes and the Social Contract Tradition* (Cambridge: Cambridge University Press, 1986).

Hampton, Jean, 'Free-rider Problems in the Production of Collective Goods', *Economics and Philosophy*, 3 (1987), 245–73.

Hanley, Ryan Patrick, *The Political Philosophy of Fénelon* (Oxford: Oxford University Press, 2020).

Hausrath, Alfred H., *Venture Simulation in War, Business, and Politics* (New York: McGraw-Hill, 1971).

Hawkey, Kate, 'A New Look At Big History', *Journal of Curriculum Studies*, 46 (2014), 163–79.

Heck, Paul van, 'In het spoor van Machiavelli: de *Politike Discoursen*, 1662, van Johan en Pieter de la Court', *LIAS*, 27 (2000), 277–318.

Hellwig, Johann Christian Ludwig, *Versuch eines aufs Schachspiel gebaueten taktischen Spiels von zwey und mehreren Personen zu Spielen* (Leipzig: S.L. Crusius, 1780).

Hellwig, Johann Christian Ludwig, *Das Kriegsspiel: ein Versuch die Wahrheit verschiedener Regeln der Kriegskunst in einem unterhaltenden Spiele anschaulich zu machen* (Braunschweig: K. Reichard, 1803).

Hendrickson, Noel, 'Counterfactual Reasoning and the Problem of Selecting Antecedent Scenarios', *Synthese*, 185 (2012), 365–86.

Hengstmengel, Joost, *Divine Providence in Early Modern Enomomic Thought* (London: Routledge, 2019).

Herberg-Rothe, Andreas, *Clausewitz's Puzzle: The Political Theory of War* (Oxford: Oxford University Press, 2007).

Hershkovitz, Shay, 'Wargame business', *Naval War College Review*, 72 (2019), 67–88.

Hilgers, Philipp von, *Kriegsspiele: Eine Geschichte der Ausnahmezustände und Unberechenbarkeiten* (München: Wilhelm Fink & Ferdinand Schöningh, 2006).

Hillenaar, Henk, *Le secret de Télémaque* (Paris: Presses Universitaires de France, 1994).

Hobbes, Thomas, *The Citizen: Philosophical Rudiments Concerning Government and Society*, in *Man and citizen. Thomas Hobbes's De Homine*, trans. by Charles T. Wood et al. (n. p. Humanities Press, 1972), 87–386.

Hobbes, Thomas, *Leviathan*, ed. Noel Malcolm, 3 vols (Oxford: Clarendon Press, 2012).

Hohrath, Daniel, 'Prolegomena zu einer Geschichte des Kriegsspiels', in Angela Giebmeyr and Helga Schnabel-Schüle, eds, '*Das Wichtigste ist der Mensch*': *Festschrift fur Klaus Gerteis zum 60. Geburtstage* (Hamburg: H.G. Schulz, 1967), 139–52.

Holborn, Hajo, 'The Prusso-German School: Moltke and the Rise of the General Staff', in Peter Paret, ed., *Makers of Modern Strategy: Military Thought from Machiavelli to the Nuclear Age* (Princeton, NJ: Princeton University Press, 1986), 281–95.

Holmes, Terence M., 'Planning versus Chaos in Clausewitz's *On War*', *The Journal of Strategic Studies*, 30 (2007), 129–51.

Holmes, Terence M., 'The Clausewitzian Fallacy of Absolute War', *The Journal of Strategic Studies*, 40 (2017), 1039–58.

Honig, Jan Willem, 'Clausewitz and the Politics of Early Modern Warfare', in Andreas Herberg-Rothe et al., eds, *Clausewitz. The State and War* (Stuttgart: Franz Steiner Verlag, 2011), 29–48.

Hont, Istvan, *Jealousy of Trade: International Competition and the Nation-State in Historical Perspective* (Cambridge, MA: Belknap Press, 2005).

Hont, Istvan, 'The Early Enlightenment Debate on Commerce and Luxury', in Mark Goldie and Robert Wokler, eds, *The Cambridge History of Eighteenth-Century Political Thought* (Cambridge: Cambridge University Press, 2006), 379–418.

Hoverbeck, C.E.B. von, *Das preussische National-Schach* (Breslau: Stadt-Buchdruckerey, 1806).

Hume, David, *A Treatise of Human Nature*, ed. L.A. Selby-Bigge and P.H. Nidditch (Oxford: Clarendon Press, 1992).

Huntington, Samuel P., 'The Clash of Civilizations', *Foreign Affairs*, (Summer 1993), 22–49.

Huntington, Samuel P., *The Clash of Civilizations and the Remaking of the World Order* (London: Simon & Schuster, 1997).

Israel, Jonathan, *The Dutch Republic: Its Rise, Greatness, and Fall* (Oxford: Clarendon Press, 1995).

Israel, Jonathan, 'The Intellectual Origins of Modern Democratic Republicanism (1660–1720)', *European Journal of Political Theory*, 3 (2004), 7–36.

Jäger, Thomas, and Rasmus Beckmann, eds, *Handbuch Kriegstheorien* (Wiesbaden: vs Verlag, 2011).

Jardine, Lisa, *Going Dutch: How England Plundered Holland's Glory* (London: HarperCollins, 2008).

Jasinski, René, 'Sur Télémaque', *A travers le XVIIe siècle*, 2 vols (Paris: Nizet, 1981), ii, 195–334.

Jaume, Lucien, 'Fénelon critique de la déraison d'État', in Yves Charles Zarka, ed., *Raison et déraison d'État: Théoriciens et théories de la raison d'État aux XVIe et XVIIe siècles* (Paris: Presses Universitaires de France, 1994), 395–422.

Jennings, Jeremy, 'The Debate about Luxury in Eighteenth- and Nineteenth-Century French Political Thought', *Journal of the History of Ideas*, 68 (2007), 79–105.

Jervis, Robert, 'Complex Systems: The Role of Interactions', in David S. Alberts and Thomas J. Czerwinski, eds, *Complexity, Global Politics, and National Security* (Washington: National Defense University, 1997), 20–31.

Johnson, Paul E., 'Simulation Modeling in Political Science', *American Behavioral Scientist*, 42 (1999), 1509–30.

Juul, Jesper, *Half-Real: Video Games between Real Rules and Fictional World* (Cambridge, MA: The MIT Press, 2011).

Kant, Immanuel, *Mutmaßlicher Anfang der Menschengeschichte*, in W. Weischedel, ed., *Schriften zur Anthropologie, Geschichtsphilosophie, Politik and Pädagogik*, 2 vols (Frankfurt am Main: Suhrkamp, 1968), 1, 83–102.

Kant, Immanuel, *Conjectures on the Beginning of Human History*, in H.S. Reis, ed., *Kant: Political Writings* (Oxford: Oxford University Press, 2013), 221–34.

Kaplan, Benjamin J., *Divided by Faith: Religious Conflict and the Practice of Toleration in Early Modern Europe* (Cambridge, MA: The Belknap Press of Harvard University Press, 2007).

Kapp, Volker, *Télémaque de Fénelon* (Tübingen: G. Narr, 1982).

Kavka, Gregory, 'Hobbes's War of All Against All', *Ethics*, 93 (1983), 291–310.

Kaye, Simon T., 'Challenging Certainty: The Utility and History of Counterfactualism', *History and Theory*, 49 (2010), 38–57.

Kellow, Geoffrey, 'Strength and Riches: Nicholas Barbon's New Politics of Commerce', *Erasmus Journal for Philosophy and Economy*, 4 (2011), 1–22.

Kessel, Eberhard, 'Zur Genesis der modernen Kriegslehre: Die Entstehungsgeschichte von Clausewitz' Buch *Vom Kriege*', *Wehrwissenschaftliche Rundschau*, 3 (1953), 405–23.

Kiesewetter, J.G.C.C., *Versuch einer faßlichen Darstellung der wichtigsten Wahrheiten der kritischen Philosophie* (Berlin: W. Oehmigke, 1803).

Kossmann, E.H., 'The Course of Dutch Political Theory in the Seventeenth Century', in *Political Thought in the Dutch Republic. Three Studies* (Amsterdam: KNAW, 2000), 25–129.

Köstlbauer, Josef, 'The Strange Attraction of Simulation: Realism, Authenticity, Virtuality', in Matthew Wilhelm Kapell and Andrew B.R. Elliott, eds, *Playing with the Past: Digital Games and the Simulation of History* (New York: Bloomsbury, 2013), 169–83.

Kr., review of Georg Venturini, *Beschreibung und Regeln eines neuen Krieges-Spiels* (Schleswig: J.G. Röhtz, 1798), in *Neue allgemeine Bibliothek: Anhang zum neun und zwanzigsten bis acht und sechzigsten Bande, enthaltend die noch übrigen Anzeigen von Büchern v. 1796–1800, Zweyte Abtheilung* (Berlin: F. Nicolai, 1803), 550–7.

Laity, Paul, *The British Peace Movement 1870–1914* (Oxford: Clarendon Press, 2001).

Le Brun, Jacques, "Notice' on Fénelon, *Explication des maximes*', in Jacques le Brun, ed., *Oeuvres*, 2 vols (Paris: Gallimard, 1983, 1997) i, 1530–45.

Lehmann, Max, *Scharnhorst*, 2 vols (Leipzig: G. Hirzel, 1886–1887).

Leistenschneider, Stephan, *Auftragstaktik im preußisch-deutschen Heer 1871–1914* (Hamburg: E.S. Mittler, 2002).

Leonhard, Robert R., 'From Operational Art to Grand Strategy', in Anthony D. McIvor, ed., *Rethinking the Principles of War* (Annapolis, MD: Naval Institute Press, 2005), 221.

Levy, Jack S., *War in the Modern Great Power System, 1495–1975* (Lexington, KY: University Press of Kentucky, 1983).

Lewis, David, *Counterfactuals* (Oxford: Blackwell, 1973).

Lightman, Bernard, 'The "History" of Victorian Scientific Naturalism: Huxley, Spencer, and the "End" of Natural History', *Studies in History and Philosophy of Biomedical Sciences*, 58 (2016), 17–23.

Little, Daniel, 'Explaining Large-Scale Historical Change', *Philosophy of the Social Sciences*, (2000), 89–112.

Lloyd, Henry, *War, Society and Enlightenment: The Works of General Lloyd*, ed. Patrick J. Speelman (Leiden: Brill, 2005).

Locke, John, *An Essay Concerning Human Understanding*, ed. Peter H. Nidditch (Oxford: Clarendon Press, 1979).

Locke, John, *Ethica A*, transcription by Paul Schuurman, *The Digital Locke Project*, http://www.digitallockeproject.nl/, accessed 3 September 2020.

Locke, John, *Of Ethic in General*, transcription by Paul Schuurman, *The Digital Locke Project*, http://www.digitallockeproject.nl/, accessed 3 September 2020.

Locke, John, *Thus I think*, transcription by Paul Schuurman, *The Digital Locke Project*, http://www.digitallockeproject.nl/, accessed 3 September 2020.

Lorson, Pierre, 'Guerre et Paix chez Fénelon', *XVIIe siècle*, (1951–1952), 207–14.

Lottes, Günther, 'Montesquieu und die Geschichte als Prozess', in Edgar Mass, ed., *Montesquieu zwischen den Disziplinen: Einzel- und kulturwissenschaftliche Zugriffe* (Berlin: Duncker & Humblot, 2010), 109–16.

Lowenthal, David, 'The design of Montesquieu's *Considerations*', *Interpretation*, 2 (1970), 144–68.

Machiavelli, *The Prince*, in *The Chief Works and Others*, 3 vols, trans. Allan Gilbert (Durham, NC: Duke University Press, 1989), i, 5–96.

Malcolm, Noel, 'Hobbes and Spinoza', in J. H. Burns, ed., *The Cambridge History of Political Thought 1450–1700* (Cambridge: Cambridge University Press, 2006), 530–57.

Malettke, Klaus, 'Fénelon, la France et le système des états Européens en 1699', in F.-X. Cuche and J. Le Brun, eds, *Fénelon: Mystique et Politique (1699–1999): Actes du colloque international de Strasbourg* (Paris: H. Champion, 2004), 469–80.

Mandeville, Bernard, *The Fable of the Bees*, ed. Philip Harth (London: Penguin, 1989).

Mason, Haydn, 'Optimism, Progress, and Philosophical History', in Mark Goldie and Robert Wokler, eds, *The Cambridge History of Eighteenth-Century Political Thought* (Cambridge: Cambridge University Press, 2006), 195–217.

Mass, Edgar, ed., *Montesquieu zwischen den Disziplinen. Einzel- und kulturwissenschaftliche Zugriffe* (Berlin: Duncker & Humblot, 2010).

Matheron, Alexandre, 'Le problème de l'évolution de Spinoza du *Traité théologico-politique* au *Traité politique*', in Edwin Curley and Pierre-François Morau, eds, *Spinoza: Issues and Directions. The Proceedings of the Chicago Spinoza Conference* (Leiden: E.J. Brill, 1990), 258–70.

Mayer, Edler von, *Eine Studie über das Kriegsspiel* (Wien: Verlag des militärwissenschaftlichen Vereines, 1874).

Mayer, Otto, 'The Origins of Feedback Control', *Scientific American*, 223 (October 1970), 110–19.

McNeill, William H., *The Pursuit of Power: Technology, Armed Force, and Society since A.D. 1000* (Chicago: University of Chicago Press, 1982).

Meckel, Klemens Wilhelm Jacob, *Studien über das Kriegsspiel* (Berlin: E.S. Mittler, 1873).

Meckel, Klemens Wilhelm Jacob, *Anleitung zum Kriegsspiele: Erster Theil: Direktiven für das Kriegsspiel* (Berlin: Vossische Buchhandlung, 1875).

Meckel, Klemens Wilhelm Jacob, *Der verbesserte Kriegsspiel-Apparat* (Berlin: Vossische Buchhandlung, 1875).

Menzies, Peter, 'Counterfactual Theories of Causation', *The Stanford Encyclopaedia of Philosophy* (Fall 2009 Edition), http://plato.stanford.edu/entries/causation-counterfactual/ accessed 24 September 2020.

Messerschmidt, Manfred, *Handbuch zur deutschen Militärgeschichte 1648–1939: Vol. iv.1. Militärgeschichte im 19. Jahrhundert 1814–1890* (München: Bernard & Graefe, 1975).

Miller, Seumas, 'Joint Action', *Philosophical Papers*, 21 (1992), 275–7.

Ministry of Defence, *Wargaming Handbook*, https://assets.publishing.service.gov.uk/government/uploads/system/uploads/attachment_data/file/641040/doctrine_uk_wargaming_handbook.pdf, accessed 1 October 2020.

M.M., *Le Jeu de la guerre, ou raffinement du jeu des echecs/Neues Kriegsspiel, oder: verbessertes Schachspiel* (Prague: F.A. Hoechenberger, 1770).

Moehler, Michael, 'Why Hobbes' State of Nature is Best Modeled by an Assurance Game', *Utilitas*, 21 (2009), 298–326.

Mokyr, Joel, *The Lever of Riches: Technological Creativity and Economic Progress* (New York: Oxford University Press, 1990).

Moltke, Helmuth von, *Memoire an Seiner Majestät den König vom 25. Juli 1868 über die bei der Bearbeitung des Feldzuges 1866 hervorgetretenen Erfahrungen*, in *Taktische-strategische Aufsätze aus den Jahren 1857 bis 1871* (Berlin: E.S. Mittler, 1900; facs. edn, London: Forgotten Books, 2018), 67–164.

Moltke, Helmuth von, *Verordnungen für die höheren Truppenführer in Großen Generalstab*, in *Taktische-strategische Aufsätze aus den Jahren 1857 bis 1871* (Berlin: E.S. Mittler, 1900; facs. edn, London: Forgotten Books, 2018), 165–215.

Montbrison, Louis Simon Joseph Bernard de, *Le Jeu de la guerre de terre et de mer, et les derniers chapitres de Tristram Shandy, trouvés dans les papiers d'Yorick* (Paris: Goujon, 1818).

Montchrestien, Antoine de, *Traicté de l'oeconomie politique*, ed. Théophile Funck-Brentano (Genève: Slatkine, 1970).

Montesquieu, *The Spirit of the Laws*, trans. Thomas Nugent, 2 vols (New York: Hafner, 1949).

Montesquieu, *De l'Esprit des lois*, in Roger Caillois, ed., *Oeuvres complètes*, 2 vols (Paris: Gallimard, 1949–1951), ii, 225–995.

Montesquieu, *De la politique*, in Roger Caillois, ed., *Oeuvres complètes*, 2 vols (Paris: Gallimard, 1949–1951), i, 112–19.

Montesquieu, *Réflexions sur le caractère de quelques princes et sur quelques événements de leur vie*, in Roger Caillois, ed., *Oeuvres complètes*, 2 vols (Paris: Gallimard, 1949–1951), i, 519–31.

Montesquieu, *Considérations sur les causes de la grandeur des Romains et de leur decadence*, eds, Françoise Weil and Cecil Courtney/*Réflexions sur la monarchie universelle en Europe*, ed. Françoise Weil, introd. Catherine Larrère and Françoise Weil, in *Oeuvres complètes de Montesquieu*, vol. 2 (Oxford: Voltaire Foundation, 2000).

Montesquieu, *Considérations sur les causes de la grandeur des Romains et de leur décadence/Réflexions sur la monarchie universelle en Europe*, ed. Cathérine Volphilhace-Auger et al. (Paris: Gallimard, 2008).

Moran, Daniel, 'Clausewitz on Waterloo: Napoleon at Bay', in Christopher Bassford, Daniel Moran and Gregory W. Pedlow, eds, *On Waterloo: Clausewitz, Wellington and the Campaign of 1815* (n. p., CLAUSEWITZ.COM, 2010).

Morris, Ian, *War, What Is It Good For? The Role of Conflict in Civilization, from Primates to Robots* (London: Profile, 2014).

Morrow, James D., *Game Theory for Political Scientists* (Princeton, NJ: Princeton University Press, 1994).

Mousnier, Roland, 'Les idées politiques de Fénelon', *XVIIe siècle*, (1951), 190–206.

Murray, M.J.R., *A History of Board Games other Than Chess* (Oxford: Oxford University Press, 1952).

Myers, Richard, 'Montesquieu on the causes of Roman greatness', *History of Political Thought*, 16 (1995), 37–47.

Nadler, Steven, 'Doctrines of explanation in late scholasticism and in the mechanical philosophy', in Daniel Garber and Michael Ayers, eds, *The Cambridge History of Seventeenth-Century Philosophy*, 2 vols (Cambridge: Cambridge University Press, 1998), ii, 513–52.

Napoléon, *Mémoires pour servir à l'histoire de France en 1815, avec le plan de la bataille de Mont-Saint-Jean* (Paris: Barrois, 1820).

Naumann, *Regiments-Kriegsspiel: Versuch einer neuen Methode des Detachments-Kriegsspiels* (Berlin: E.S. Mittler, 1877).

Neumann, *Directiven für das Festungs-Kriegsspiel* (Berlin: Vossische Buchhandlung, 1872).

Neumann, J. von, and O. Morgenstern, *The Theory of Games and Economic Behavior* (Princeton, NJ: Princeton University Press, 1944).

Niderst, Alain, 'Le Quiétisme de *Télémaque*?', in F.-X. Cuche and J. Le Brun, eds, *Fénelon: Mystique et Politique (1699–1999)*, Actes du colloque international de Strasbourg (Paris: H. Champion, 2004), 205–16.

Nohr, Rolf F., and Serjoscha Wiemer, 'Strategie Spielen', in Rolf F. Nohr and Serjoscha Wiemer, eds, *Strategie Spielen* (Münster: Lit Verlag, 2008), 7–28.

Nohr, Rolf F., and Stefan Böhme, *Johann C.L. Hellwig und das Braunschweiger Kriegsspiel* (Braunschweig: Appelhans-Verlag, 2009).

Nolan, Daniel, 'Why Historians (and Everyone Else), Should Care about Counterfactuals', *Philosophical Studies*, 163 (2013), 317–35.

North, Douglass C., and Robert Paul Thomas, *The Rise of the Western World* (Cambridge: Cambridge University Press, 1973).

Nyden-Bullock, Tammy, 'Radical Cartesian Politics: Velthuysen, De la Court, and Spinoza', *Studia Spinozana*, 15 (1999), 35–65.

Oake, Roger B., 'Montesquieu's Analysis of Roman History', *Journal of the History of Ideas*, 16 (1955), 44–59.

Oetting, Dirk W., *Auftragstaktik: Geschichte und Gegenwart einer Führungskonzeption* (Bonn: Report Verlag, 1993).

Offer, John, 'Social Change and Selectionist Thought: On Spencer, Darwin and Runciman', *The Sociological Review*, 58 (2010), 305–26.

Olson, Mancur, *The Logic of Collective Action: Public Goods and the Theory of Groups* (Cambridge, MA: Harvard University Press, 1971).

Opiz, Johann Ferdinand, *Das Opiz'sche Kriegsspiel, ein Beitrag zur Bildung künftiger und zur Unterhaltung selbst der erfahrensten Taktiker* (Halle: Hendels Verlag, 1806).

Oswalt, Ivar, 'Current Applications, Trends, and Organizations in U.S. Military Simulation and Gaming', *Simulation & Gaming*, 24 (1993), 153–89.

Palmer, R.R., 'Frederick the Great, Guibert, Bülow: From Dynastic to National War', in Peter Paret, ed., *Makers of Modern Strategy: Military Thought from Machiavelli to the Nuclear Age* (Princeton, NJ: Princeton University Press, 1986), 91–119.

Paret, Peter, ed., *Makers of Modern Strategy: Military Thought from Machiavelli to the Nuclear Age* (Princeton, NJ: Princeton University Press, 1986).

Paret, Peter, *Clausewitz and the State: The Man, His Theories and His Times* (Princeton, NJ: Princeton University Press, 2007).

Paret, Peter, *The Cognitive Challenge of War: Prussia 1806* (Princeton, NJ: Princeton University Press, 2009).

Parker, Geoffrey, *The Military Revolution: Military Innovation and the Rise of the West, 1500–1800* (Cambridge: Cambridge University Press, 1996).

Parker, Geoffrey, 'States Make War but Wars also Break States', *The Journal of Military History*, 74 (2010), 11–34.

Parker, Geoffrey, *Global Crisis: War, Climate Change and Catastrophe in the Seventeenth Century* (New Haven: Yale University Press, 2013).

Pearce, Trevor, 'From "Circumstances" to "Environment": Herbert Spencer and the Origins of the Idea of Organism—Environment Interaction', *Studies in History and Philosophy of Biological and Biomedical Sciences*, 41 (2010), 241–52.

Penny, David, 'Epigenetics, Darwin, and Lamarck', *Genome Biology and Evolution*, 7 (2015), 1758–60.

Perinetti, Dario, 'Philosophical Reflection on History', in Knud Haakonssen, ed., *The Cambridge History of Eighteenth-Century Philosophy*, 2 vols (Cambridge: Cambridge University Press, 2006) ii, 1127–8.

Perkuhn, Johann Gottlieb, *Beschreibung eines Kriegsspiels zum Gebrauch für Militairs* (Hamburg: Appel, 1817).

Perla, Peter P., *The Art of Wargaming* (Annapolis, Maryland: U.S. Naval Institute, 1990).

Peterson, Jon, 'A Game Out of All Proportions: How a Hobby Miniaturized War', in Pat Harrigan and Matthew G. Kirschenbaum, eds, *Zones of Control: Perspectives on Wargaming* (Cambridge, MA: The MIT Press, 2016), 3–31.

Pias, Claus, *Computer—Spiel—Welten* (München: Sequenzia, 2010).

Postigliola, Alberto, ed., *Storia e ragione. Les Considérations sur les causes de la grandeur des Romains et de leur décadence di Montesquieu nel 250° della publicatione* (Naples: Liguori, 1987).

Price, J.L., *Holland and the Dutch Republic in the Seventeenth Century: The Politics of Particularism* (Oxford: Clarendon Press, 1994).

Racault, Jean-Michel, *L'Utopie narrative en France et en Angleterre 1675–1761*, Studies on Voltaire and the Eighteenth Century, vol. 280 (Oxford: The Voltaire Foundation at the Taylor Institution, 1991).

Rahe, Paul A., 'The book that never was', *History of Political Thought*, 26 (2005), 43–89.

Rahe, Paul A., *Montesquieu and the Logic of Liberty* (New Haven: Yale University Press, 2009).

Rahe, Paul A., *Soft Despotism, Democracy's Drift: Montesquieu, Rousseau, Tocqueville & The Modern Prospect* (New Haven: Yale University Press, 2009).

Rasmusen, Eric, *Games and Information: An Introduction to Game Theory* (Malden, MA: Blackwell, 2007).

Regent, Nikola, 'In the Shadow of Lacedaemon: Luxury, Wealth and Early-Modern Republican Thought', *Journal of the History of Economic Thought*, 41 (2019), 477–509.

Reinhard, Wolfgang, and Heinz Schilling, eds, *Die katholische Konfessionalisierung* (Gütersloh: Gütersloher Verlagshaus, 1995).

Reiss, Julian, 'Counterfactuals, Thought Experiments, and Singular Causal Analysis in History', *Philososophy of Science*, 76 (2009), 712–23.

Reiswitz (=Reisswitz) [Georg Heinrich Rudolf Johann von], *Anleitung zur Darstellung militairischer Manöver mit dem Apparat des Kriegs-Spiels* (Berlin: Trowitzsch, 1824).

Reiswitz, Georg Leopold von, *Taktisches Kriegsspiels oder Anleitung zu einer mechanischen Vorrichtung um taktische Manoeuvres sinnlich darzustellen* (Berlin: Gebrüdern Güdicke, 1812).

Renwick, Chris, 'The Practice of Spencerian Science: Patrick Gedde's Biosocial Program, 1876–1889', *Isis*, 100 (2009), 36–57.

Richardson, Kurt A., 'Systems and Complexity: Part 1', *E:CO*, 6 (2004), 75–9.

Richet, Melvin, *The Political Theory of Montesquieu* (Cambridge: Cambridge University Press, 1977).

Riley, Patrick, 'Rousseau, Fénelon, and the Quarrel Between the Ancients and the Moderns', in Patrick Riley, ed., *The Cambridge Companion to Rousseau* (Cambridge: Cambridge University Press, 2001), 87–93.

Rogers, Clifford J., 'Clausewitz, Genius, and the Rules', *The Journal of Military History*, 66 (2002), 1167–76.

Rosenberg, Daniel, 'Joseph de Maistre on War and Peace: Ritual and Ritualism', *The Philosophical Journal of Conflict and Violence*, 3 (2019), 35–51.

Rosenfield, Gavriel, 'Why Do We Ask "What if?" Reflections on the Function of Alternate History', *History and Theory*, 41 (2002), 90–103.

Rosenthal, Michael A., 'Two Collective Action Problems in Spinoza's Contract Theory', *History of Philosophy Quarterly*, 15 (1998), 389–409.

Ross, Don, 'Game Theory', in *Stanford Encyclopedia of Philosophy*, https://plato.stanford.edu/entries/game-theory/, accessed 23 October 2020.

Roth, Paul A., and Thomas A. Ryckman, 'Chaos, Clio, and Scientific Illusions of Understanding', *History and Theory*, 34 (1995), 30–44.

Rothkrug, Lionel, *Opposition to Louis XIV: The Political and Social Origins of the French Enlightenment* (Princeton, NJ: Princeton University Press, 1965).

Rueger, Alexander, 'Connection and Influence: A Process Theory of Causation', *Journal for General Philosophy of Science*, 37 (2006), 77–97.

Sabin, Philip, *Simulating War: Studying Conflict through Simulation Games* (2nd ed., London: Bloomsbury, 2014).

Saint-Simon, Claude-Henri de, *Du système industriel* (Paris: Antoine-Augustin Renouard, 1821).

Sandkühler, Gunnar, 'Das Kriegsspiel als Erziehungsmodell', in Rolf F. Nohr and Stefan Böhme, eds, *Johann C.L. Hellwig und das Braunschweiger Kriegsspiel. Die Auftritte des Krieges Sinnlich machen* (Braunschweig: Appelhans-Verlag, 2009), 28–37.

Sandkühler, Gunnar, 'Die philantropische Versinnlichung Hellwigs Kriegsspiel als pädagogisches und immersives Erziehungsmodell', in Rolf F. Nohr and Serjoscha Wiemer, eds, *Strategie Spielen* (Münster: Lit Verlag, 2008), 69–86.

Sayre, Farrand, *Map Maneuvres and Tactical Rides* (Fort Leavenworth, KS: The Army Service Schools Press, n. d.).

Scharnhorst, Gerhard von, *Über die Wirkung des Feuergewehrs* (Berlin: G.E. Nauck, 1813).

Scharnhorst, Gerhard von, *Private und dienstliche Schriften*, vol. 3: *Lehrer, Artillerist, Wegbreiter*, ed., Johannes Kunisch (Köln: Böhlau, 2002).

Schering, Walter Malmsten, *Wehrphilosophie* (Leipzig: J. A. Barth, 1939).

Schlegel, Klaus, 'General der Infanterie Dr. phil. h.c. Julius von Verdy du Vernois', in Helmut Damerau, ed., *Deutsches Soldatenjahrbuch 1982: Dreissigster deutscher Soldatenkalender* (München: Schildverlag, 1982), 71–5.

Schöffer, Ivo, 'Did Holland's Golden Age Coincide with a Period of Crisis?', in Geoffrey Parker, ed., *The General Crisis of the Seventeenth Century* (London: Routledge & Kegan Paul, 1978), 78–109.

Schuurman, Paul, 'René Descartes' hybride fysica: zekere principes en hypothetische modellen', *Tijdschrift voor Filosofie*, 63 (2001), 261–81.

Schuurman, Paul, 'Fénelon on Luxury, War and Trade in the *Telemachus*', *History of European Ideas*, 38 (2012), 1–21.

Schuurman, Paul, 'Determinism and Causal Feedback Loops in Montesquieu's Explanation for the Military Rise and Fall of Rome', *British Journal for the History of Philosophy*, 21 (2013), 507–28.

Schuurman, Paul, 'The Cat's Grand Strategy: Pieter de la Court (1618–1685) on Holland and the Challenges and Prospects of Free-Riding Behaviour During the General Crisis of the Seventeenth Century', *History of European Ideas*, 40 (2014), 1–19.

Schuurman, Paul, 'Clausewitz on real war', *Peace Review*, 26 (2014), 372–9.

Schuurman, Paul, 'War as a System: A Three-Stage Model for the Development of Clausewitz's Thinking on Military Conflict and Its Constraints', *Journal of Strategic Studies*, 37 (2014), 926–48.

Schuurman, Paul, 'Herbert Spencer and the Paradox of War', *Intellectual History Review*, 26 (2016), 519–35.

Schuurman, Paul, 'Kant on War and the Idea of Progress', *Twijfel*, (October 2016), 6–8.

Schuurman, Paul, 'Models of War 1770–1830: The Birth of Wargames and the Trade-off Between Realism and Simplicity', *History of European Ideas*, 43 (2017), 442–5.

Schuurman, Paul, 'What-If At Waterloo: Carl von Clausewitz's Use of Historical Counterfactuals in his History of the Campaign of 1815', *Journal of Strategic Studies*, 40 (2017), 1016–38.

Schuurman, Paul, 'A Game of Contexts: Prussian-German Professional Wargames and the Leadership Concepts of Mission Tactics 1870–1880', *War in History*, 28 (2021), 504–24.

Schwartz, Karl, *Leben des Generals Carl von Clausewitz und der Frau Marie von Clausewitz geb. Gräfin von Brühl*, 2 vols (Berlin: F. Dümmler, 1878).

Schwendimann, Beat, 'What is the Difference Between a Simulation and a Model?', http://proto-knowledge.blogspot.nl/2010/12/what-is-difference-between-simulation.html, accessed 8 November 2016.

Sekora, John, *Luxury: The Concept in Western Thought, Eden to Smolett* (Baltimore: The Johns Hopkins University Press, 1977).

Senarclens, Vanessa de, 'Montesquieus historische Herangehensweise am Beispiel Roms', in Edgar Mass, ed., *Montesquieu zwischen den Disziplinen: Einzel- und kulturwissenschaftliche Zugriffe* (Berlin: Duncker & Humblot, 2010), 125–34.

Shackleton, Robert, *Montesquieu: A Criticial Biography* (Oxford: Oxford University Press, 1961).

Shaw, Leslie, 'Fénelon and Christian Agrarianism', https://www.academia.edu/8371021/Fénelon_and_Christian_Agrarianism, accessed 29 October 2020.

Shklar, Judith, *Men and Citizens: A Study of Rousseau's Social Theory* (Cambridge: Cambridge University Press, 1985).

Sigg, Marco, *Der Unterfüher als Feldherr im Taschenformat: Theorie und Praxis der Auftragstaktik im deutschen Heer 1869–1945* (Paderborn: Ferdinand Schöningh, 2014).

Silberner, Edmund, *La guerre et la paix dans l'histoire des doctrines économiques* (Paris: Sirey, 1957).

Skinner, Michael K., 'Environmental Epigenetics and a Unified Theory of the Molecular Aspects of Evolution: A Neo-Lamarckian Concept that Facilitates Neo-Darwinian Evolution', *Genome Biology and Evolution*, 7 (2015), 1296–302.

Sleigh, Robert, et al., 'Determinism and Human Freedom', in Daniel Garber and Michael Ayers, eds, *The Cambridge History of Seventeenth-Century Philosophy*, 2 vols (Cambridge: Cambridge University Press, 1998), ii, 1195–278.

Snelder, H.A.M., 'Drebbel, Cornelis Jacobsz', in Wiep van Bunge et al., eds, *The Dictionary of Seventeenth and Eighteenth-Century Dutch Philosophers*, 2 vols (Bristol: Thoemmes Press, 2003), i, 273–5.

Soll, Jacob, 'Accounting for Government: Holland and the Rise of Political Economy in Seventeenth-Century Europe', *Journal of Interdisciplinary History*, 40 (2009), 215–38.

Sonenscher, Michael, *Before the Deluge: Public Debt, Inequality, and the Intellectual Origins of the French Revolution* (Princeton, NJ: Princeton University Press, 2007).

Spencer, Herbert, *The Principles of Biology*, 2 vols (London: Williams and Norgate, 1864, 1867).

Spencer, Herbert, *The Social Organism*, in *Essays: Scientific, Political, & Speculative* (London: G. Norman, 1891), 272–6, http://www.gutenberg.org/files/29869/29869-h/29869-h.htm, accessed 30 June 2015.

Spencer, Herbert, 'The Relations of Biology, Psychology, and Sociology', *Popular Science Monthly*, 50 (1896), 163–71, https://en.wikisource.org/wiki/Popular_Science_Monthly/Volume_50/December_1896/The_Relations_of_Biology,_Psychology,_and_Physiology, accessed 30 June 2015.

Spencer, Herbert, *First Principles* (London: Williams & Norgate, 1904).

Spencer, Herbert, *The Man versus The State*, in *Political Writings*. (Cambridge: Cambridge University Press, 1994), 59–175.

Spencer, Herbert, *The Principles of Sociology in three volumes*. (New Brunswick: Transaction, 2010).

Spier, Fred, 'Big History: The Emergence of a Novel Interdisciplinary Approach', *Interdisciplinary Science Reviews*, 33 (2008), 141–7.

Spier, Fred, *Big History and the Future of Humanity* (Chichester: J. Wiley, 2015).

Spinoza, Benedictus de, 'Letters: January 1671–Late 1676', in *The Collected Works of Spinoza*, 2 vols, trans. Edwin Curley (Princeton, NJ: Princeton University Press, 2016), ii, 374–490.

Spinoza, Benedictus de, 'Theological-Political Treatise', in *The Collected Works of Spinoza*, 2 vols, trans. Edwin Curley (Princeton, NJ: Princeton University Press, 2016), ii, 65–356.

Stoker, Donald, *Clausewitz: His Life and Work* (Oxford: Oxford University Press, 2014).

Stokes Brown, Cynthia, *Big History: From the Big Bang to the Present* (London: The New Press, 2007).

Strachan, Hew, *Clausewitz's On War* (New York: Grove Press, 2007).

Strosetzki, Christoph, 'Die "Dekadenz" Roms, ihre Ursachen und ihre Dialektik bei Montesquieu', in Edgar Mass, ed., *Montesquieu zwischen den Disziplinen: Einzel- und kulturwissenschaftliche Zugriffe* (Berlin: Duncker & Humblot, 2010), 71–88.

Stübing, Heinz, *Scharnhorst: Die Reform des preußischen Heeres* (Göttingen: Muster-Schmidt, 1988).

Sumida, Jon Tetsuro, *Decoding Clausewitz: A New Approach to On War* (Lawrence, KS: University Press of Kansas, 2008).

Thomson, J.A., *Herbert Spencer* (London: Dent, 1906).

Tijn, Th. van, 'Pieter de la Court: een buitenbeen', in H.W. Blom and I.W. Wildenberg, eds, *Pieter de la Court en zijn tijd* (Amsterdam: Holland Universiteits Pers, 1986), xi–xiv.

Troschke, Theodor, 'Zum Kriegsspiel', *Militär-Wochenblatt*, 35 (1869), 35 and 276–7.

Trotha, Lebrecht Ernst Michael Thilo von, *Anleitung zum Gebrauch des Kriegsspiel-Apparates zur Darstellung von Gefechtsbildern mit Berücksichtigung der Wirkung der jetzt gebräuchlichen Waffen* (Berlin: E. S. Mittler, 1870).

Trotha, Lebrecht Ernst Michael Thilo von, review of Julius Verdy du Vernois, *Beitrag zum Kriegsspiel* (Berlin: E.G. Mittler, 1876), in *Militair-Wochenblatt* (1876), 6, 102–7.

Trousson, Raymond, *Voyages au pays de nulle part: Histoire littéraire de la pensée utopique* (Brussels: Éditions de l'Université de Bruxelles, 1975).

Tschischwitz, W. von, *Anleitung zum Kriegs-Spiel* (4th ed., Neisse: J. Graveur, 1874).

Tsouras, Peter S., *Napoleon Victorious: An Alternate History of the Battle of Waterloo* (Barnsley: Greenhill Books, 2018).

Tuomela, Raimo, 'On the Structural Aspects of Collective Action and Free-Riding', *Theory and Decision*, 32 (1992), 165–202.

Turner, Jonathan H., *Herbert Spencer: A Renewed Appreciation*. (New Delhi: Sage Publications, 1985).

Turner, Jonathan H., and Seth Abrutyn, 'Returning the "Social" to Evolutionary Sociology: Reconsidering Spencer, Durkheim, and Marx's Models of "Natural" Selection', *Sociological Perspectives*, 60 (2017), 527–66.

Türpe, Andrée, 'Dialektisches Denken in den Auffassungen von Clausewitz', *Deutsche Zeitschrift für Philosophie*, 28 (1980), 709–18.

Uffindell, Andrew, 'Napoleon at Waterloo', in Jonathan North, ed., *The Napoleon Options: Alternate Decisions of the Napoleonic Wars* (London: Greenhill, 2000), 187–202.

Uyl, Douglas J. den, *Power, State and Freedom: An Interpretation of Spinoza's Political Philosophy* (Assen: Van Gorcum, 1983).

Vego, Milan N., *Joint Operational Warfare: Theory and Practice* (Newport, RI: U.S. Naval War College, 2009), 73–95.

Velema, Wyger, 'That a Republic is Better than a Monarchy', in Martin van Gelderen and Quentin Skinner, eds, *Republicanism: a Shared European Heritage*, 2 vols (Cambridge: Cambridge University Press, 2002), i, 9–25.

Venturini, Georg, *Beschreibung und Regeln eines neuen Krieges-Spiels* (Schleswig: J.G. Röhtz, 1798).

Venturini, Georg, *Darstellung eines neuen Kriegsspiels* (Leipzig: Johann Conrad Hinrichs, 1804).

Verdy du Vernois, Julius, *Beitrag zum Kriegsspiel: Mit einem Plane* (Berlin: E.S. Mittler, 1876).

Verdy du Vernois, Julius, *Studien über Truppenführung* (Berlin: E.S. Mittler, 1898).

Volpilhac-Auger, Catherine, 'Ex oriente nox? Le paradoxe byzantin chez Montesquieu', *Dix-huitième Siècle*, 35 (2003), 393–404.

Volpilhac-Auger, Catherine, 'Montesquieu et l'histoire: une occasion manquée?', in Edgar Mass, ed., *Montesquieu zwischen den Disziplinen: Einzel- und kulturwissenschaftliche Zugriffe* (Berlin: Duncker & Humblot, 2010), 135–46.

Vries, Jan de, 'The Economic Crisis of the Seventeenth Century after Fifty Years', *Journal of Interdisciplinary History*, 40 (2009), 151–94.

Vuia, Octavian, *Montesquieu und die Philosophie der Geschichte*. (Frankfurt am Main: P. Lang, 1998).

Wahnbeck, Till, *Luxury and Public Happiness: Political Economy in the Italian Enlightenment* (Oxford: Clarendon Press, 2004).

Walter, Dierk, *Preußische Heeresreformen 1807–1870: Militärische Innovation und der Mythos der "Roonschen Reform"* (Paderborn: F. Schöningh, 2003).

Wattles, Jeffrey, 'Teleology Past and Present', *Zygon*, 4 (2006), 445–464.

Watts, Barry D., *Clausewitzian Friction and Future War* (Washington, DC: Institute for National Strategic Studies, 2004).

Weikart, Richard, 'Was Darwin or Spencer the Father of laissez-faire Social Darwinism?', *Journal of Economic Behavior & Organization*, 71 (2009), 20–8.

Wellington, [Arthur Wellesley, Duke of], *Memorandum on the Battle of Waterloo*, in Christopher Bassford, Daniel Moran and Gregory W. Pedlow, eds and trans., *On Waterloo: Clausewitz, Wellington and the Campaign of 1815* (n. p., CLAUSEWITZ.COM, 2010), 219–35.

Wellstead, Peter, and Mark Read, 'Engine Speed Control', http://www.control-systems-principles.co.uk/whitepapers/engine-speed-control.pdf, accessed 15 September 2020.

Welzer, Harald, *Climate Wars: Why People Will be Killed in the Twenty-First Century* (Cambridge: Polity Press, 2019).

Weststeijn, Arthur, 'From the Passion of Self-Love to the Virtue of Self-Interest: the Republican Morals of the Brothers De la Court', *European Review of History*, 17 (2010), 75–92.

Weststeijn, Arthur, *Commercial Republicanism in the Dutch Golden Age: The Political Thought of Johan & Pieter de la Court* (London: Brill, 2012).

Wette, Wolfram, *Militarismus in Deutschland: Geschichte einer kriegerischen Kultur* (Frankfurt am Main: Fisher Taschenbuch, 2011).

Wildenberg, Ivo W., *Johan & Pieter de la Court (1622–1660 & 1618–1685)* (Amsterdam: Holland Universiteits Pers, 1986).

Wittmann, Jochen, *Auftragstaktik—Just a Command Technique or the Core Pillar of Mastering the Military Operational Art?* (Norderstedt: Books on Demand, 2012).

Young, John P., *A Survey of Historical Development in War Games* (Bethseda: Operations Research Office, 1959).

Zakariya, Nasser, 'Is History Still a Fraud?', *Historical Studies in the Natural Sciences*, 43 (2013), 631–41.

Zammito, John, 'Teleology Then and Now: The Question of Kant's Relevance for Contemporary Controversies over Function in Biology', *Studies in History and Philosophy of Biological and Biomedical Sciences*, 37 (2006), 748–70.

Zhang, David D., et al., 'Global Climate Change, War, and Population Decline in Recent Human History', *Proceedings of the National Academy of Sciences*, 104 (2007), 19214–9.

Zipser, Carl, *Anleitung zur Darstellung militärischer Manöver mit Hilfe des Kriegsspiel-Apparates, bei Zugrundelegung der neuesten taktischen Normen* (2nd ed., Josefstadt: C. Hoser Trautenau, 1876).

Zychlinski, Franz von, *Geschichte des 24sten Infanterie-Regiments*, ii: *1816–1838* (Berlin: Mittler, 1857).

Index

www.ingramcontent.com/pod-product-compliance
Lightning Source LLC
LaVergne TN
LVHW010626300926
849162LV00001B/79